The identification of flowering plant families

Third edition

The identification of flowering plant families

including a key to those native and
cultivated in north temperate regions

P. H. DAVIS D.SC.
FORMERLY PROFESSOR OF BOTANY IN THE
UNIVERSITY OF EDINBURGH

J. CULLEN D.SC.
ASSISTANT KEEPER, ROYAL BOTANIC GARDEN, EDINBURGH

THIRD EDITION
completely revised and edited by J. Cullen

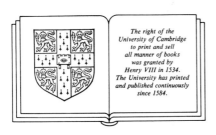

The right of the
University of Cambridge
to print and sell
all manner of books
was granted by
Henry VIII in 1534.
The University has printed
and published continuously
since 1584.

CAMBRIDGE UNIVERSITY PRESS
Cambridge
New York Port Chester
Melbourne Sydney

Published by the Press Syndicate of the University of Cambridge
The Pitt Building, Trumpington Street, Cambridge CB2 1RP
40 West 20th Street, New York NY 10011, USA
10 Stamford Road, Oakleigh, Melbourne 3166, Australia

First published by Oliver and Boyd 1965
Second edition published by Cambridge University Press 197ᶜ
Reprinted 1980
Third edition 1989

Printed in Great Britain at the University Press, Cambridge

British Library cataloguing in publication data

Davis, P. H. (Peter Hadland), 1918– .
The identification of flowering plant families. – 3rd ed.
1. Flowering plants. Northern Hemisphere.
Temperate regions. Identification manuals.
I. Title. II. Cullen, J. (James), 1936–
582.13'09181'3

Library of Congress cataloguing in publication data

Davis, P. H. (Peter Hadland).
The identification of flowering plant families, including
a key to those native and cultivated in north temperate regions /
P. H. Davis, J. Cullen – 3rd ed. / completely rev. and edited by J. Cullen.
    p.   cm.
Bibliography: p.
Includes index.
ISBN 0 521 37335 2
ISBN 0 521 37707 2 (pbk.)
1. Angiosperms – Identification. 2. Flowers – Identification.
I. Cullen. J. (James). II. Title.
QK495.A1D38   1989.
582.13'09123–dc19

ISBN 0 521 37335 2 hard covers
ISBN 0 521 37707 2 paperback
(Second edition ISBN 0 521 22111 0 hard covers
                ISBN 0 521 29359 6 paperback)

Contents

Illustrations

Preface

That a book of this type should appear in a third edition implies that it has a certain usefulness and practicability. The main purpose of the changes made to the text in preparing this edition has been to enhance and improve these qualities.

Since 1979, when the second edition was published, numerous changes have taken place in taxonomic thinking and in public attitudes to plants and wildlife generally. Some of these changes continue the trends noted between the first (1965) and second editions, particularly the continuing decline of the teaching of plant taxonomy in the Universities and the increasing separation of professional plant taxonomists from the public that need the results of their work.

On the other hand, there have been some encouraging developments as well, especially the increasing public awareness of the need for conservation, which has led to a renewed interest in plants (and animals). This interest is in part satisfied by excellent wildlife programmes on television and by the numerous, well-illustrated popular books on plants and animals available today. Less, however, is done for those who have a more serious, practical interest, and wish to know more about the scientific background. This edition is intended for such people, whether they are botanists, gardeners, landscape architects or otherwise involved with plants, either as students, professionally or as a hobby.

The main changes from the second edition are as follows:

(a) The adoption of the Engler & Prantl taxonomic system as the basis for the recognition of families and their taxonomic sequence. This may appear regressive to professional taxonomists, as many more up-to-date systems have been published since 1979. However, the Engler & Prantl system still seems to be the best one for general purposes: it is well documented and widely used; its use in *The European Garden Flora* (1984 and continuing) gives further impetus for its use here.

(b) The keys have been modified to take account of errors and difficulties in use that have come to notice since 1979, and the section 'Using the keys' has been enlarged to provide more guidance.

(c) The section 'Further identification' has been completely re-written.

(d) Many minor changes, especially in terminology, have been incorporated.

Finally, the last paragraph of the preface to the second edition still remains valid, and I quote it without apology: 'In our view a natural classification of plants and their correct identification (for which the families are an important step on the road) remain essential for the progress of biology on a broad front.'

James Cullen

Edinburgh
July 1988

Acknowledgements

Over the years of this book's existence, many botanists and others have helped with advice and information on matters of detail. These are now too numerous to mention individually, and include several correspondents who have pointed out errors or omissions. I hope they will accept this general acknowledgement. In the preparation of this third edition I am particularly grateful to Dr J. C. M. Alexander, Dr G. C. G. Argent and Miss Suzanne Maxwell for both botanical and general assistance. The illustrations, reprinted from the second edition, are by Miss R. M. Smith.

James Cullen

Abbreviations

Mostly used in the short descriptions of the families on pp. 73–114.

A	androecium (stamens)	*N*	North
act	radially symmetric (actinomorphic)	n	many
		opp	opposite
alt	alternate	ov	ovules
antipet	antipetalous	P	perianth
ax	axile placentation	par	parietal placentation
bisex	bisexual	perig	perigynous
C	corolla	rar	rarely
C	Central	*S*	South
E	East	solit	solitary
epig	epigynous	stip	stipules, stipulate
exstip	without stipules	sup	superior
fl	flower	*Temp*	temperate
fr	fruit	*Trop*	tropical
G	gynoecium (ovary)	unisex	unisexual
hypog	hypogynous	usu	usually
inf	inferior	var	various
infl	inflorescence	*W*	West
Is	Islands	zyg	bilaterally symmetric (zygomorphic)
K	calyx		
lvs	leaves	/	or

Introduction

The identification of the family to which a plant belongs is usually the first step in its complete identification. This key attempts to provide a means of identification for all Angiosperm (flowering plant) families native or cultivated, out-of-doors or under glass, in north temperate regions. In practice, we have taken the southern limit of our area as approximately 30° N, thus excluding all of Mexico and Florida in the New World and most of India and subtropical China in the Old World. A few, mainly tropical families with a small number of genera native in China north of this limit have also been excluded, as, unless cultivated in Europe or North America, they are so infrequently seen. The key covers 285 families.

As far as cultivated plants are concerned, a few tropical families which rarely flower in cultivation have been excluded. The key has been constructed to allow for the identification of the frequently cultivated representatives of tropical and southern hemisphere families; it may not work for other genera. No attempt has been made to cope with the double-flowered, wilder excesses of the plant breeder.

The family is merely one level of the taxonomic hierarchy, but it is generally the highest level that has any importance in identification (but see p. 26 for details of the Monocotyledones and Dicotyledones). Families contain genera, which themselves contain species. A family may contain effectively only a single species, e.g. Adoxaceae, which comprises the single genus *Adoxa*, itself made up of the single species *A. moschatellina*. On the other hand, large families contain some hundreds of genera and several thousand species (e.g. Orchidaceae, Compositae, Leguminosae). This variation in size means that some families are considerably more variable than others, and that characters that are generally diagnostic in some cases are not so in others; hence many families are keyed out more than once in this key. The variability also means that what constitutes a family is to a considerable extent a matter of opinion, and in recent years there has been a strong tendency to split up the large families into segregates. In this book we have attempted to take a middle way between the 'lumpers' and 'splitters', following a well-established taxonomic system (that of Engler & Prantl as expressed in *Syllabus der Pflanzenfamilien*, edn 12, volume 2, ed. by H. Melchior), while indicating, by means of sub-keys under the brief descriptions of the families (pp. 73–114), some of the more widely recognised segregates.

Families are grouped together in terms of what are considered to be their affinities or relationships, into Orders (whose names are similar to those of the families, but end in '-ales'). These are briefly mentioned with the descriptions of the families.

An arrangement of families in their orders into a particular linear sequence forms a taxonomic system. The sequence of families is used by authors of systems as a means of indicating their ideas about how the Angiosperms have evolved. There are many differing systems available in the taxonomic literature, though none of them has won particularly wide acceptance (the differences between them are often trivial). That followed here is the most generally useful for identification purposes, being widely followed in Floras (e.g. *Flora Europaea*, *The European Garden Flora*), well-documented and widely available. Its adoption here is for convenience and practicality and implies no acceptance of any particular view of the evolution of the Angiosperms.

The short descriptions of the families are intended both as a check on identification and as a terse presentation of the important family characteristics. These descriptions refer to the families as wholes, not only to those representatives catered for in the key. The distribution of each family has been given, though necessarily without great detail. For families consisting of only a single genus, the name of that genus is also given.

To make the book as useful as possible, we have provided a glossary; this is mainly designed to assist with understanding of the descriptions of the families, where conciseness was a major aim. In the keys themselves, fewer technical terms are used. Terms requiring a more extensive explanation than is possible in a glossary are discussed in the section entitled 'Usage of terms' (p. 4). For users not particularly experienced in plant identification, a section on 'Examining the plant' is also included. This details many of the important characteristics easily overlooked by a too superficial examination of specimens, and should help to avoid 'howlers' in identification.

The short section entitled 'Further identification and annotated bibliography' (p. 115) is intended to help the user to proceed further with the identification process. For reasons of space this section is short, and refers to major works and lists only.

It is not possible, in a book of this size, to give information about the theory and practice of plant classification. It must be remembered that the name of a plant is only a key to information about it, not an end in itself. Any user whose interest in this aspect of the subject is stimulated by the practice of identifying plants is recommended to consult a good library where botanical books are well represented.

Usage of terms

A few terms need a fuller explanation than can be given in a glossary; these are the terms referring to the relative positions of the floral organs, placentation and aestivation, which are discussed and illustrated below.

It must be emphasised that in this section we have not so much attempted to define terms as to explain how they are used in this book. This has been necessary because there is little consistency in their usage in the botanical literature. We have avoided, as much as possible, explanations which require the acceptance of particular theories as to the origins, homologies, etc., of the structures, and have contented ourselves with a purely descriptive terminology which should be practicable and widely applicable whatever interpretation is put on the nature of the floral parts.

Hypogyny, perigyny and epigyny

Two sets of terms are used in the description of the relative positions of the floral organs. One (superior/inferior) is generally used with reference to the position of the ovary with respect to the other organs of the flower; the other (hypogyny/perigyny/epigyny) appears to refer to the position of the other organs with respect to the ovary, but is perhaps best expressed in terms of the apparent fusion of organs of different whorls, rather than by reference to ovary position alone. The phenomena covered by both sets of terms are best seen in longitudinal sections of flowers (see figs. 1–4).

Egler (*Chronica Botanica* 12 (1951), 169–73) has discussed both these sets of terms and has recommended their replacement. We have not followed his recommendation because our aim has been to refine and clarify the standard terminology (which is used in different senses by different authors) rather than to try to replace it.

The terms referring to ovary position are not especially ambiguous – a superior ovary is one borne on the receptacle (torus – the top of the flower-stalk) above the insertion of the other floral organs (regardless of whether these are free from each other or variously united, see figs. 2 and 3(a)–(c); an inferior ovary is one borne below the point of insertion of the other floral organs, so that they appear to be borne on or near the top of the ovary (see figs. 3(d), (e); 4(a), (b)). The only difficulty is introduced by the occasional occurrence of an intermediate condition in which the ovary is said to be half-inferior. In this condition the tops of the cells (loculi) of the ovary occur above the point of insertion of the other floral organs (as seen in a longitudinal section of the flower) and the bottoms of the cells below this point (see fig. 3(d)).

The other terminology is more difficult to apply. One of the main problems with these terms is that they have frequently been applied to the flower as a whole. We have found this practice to be misleading, and prefer to use the terms with reference to the perianth and stamens. We hope that this usage, which is based on that formulated by De Candolle in his *Theorie Elementaire de Botanique* (1813), will be accepted as providing a less ambiguous description of the relationships of the floral parts.

Table 1 and the accompanying diagram (fig. 1) should make clear the usage of the terms that we have adopted and show the relationship between the two sets of terms discussed above.

A few words of explanation may help to make the table easier to understand. The 'ring or collar of tissue' mentioned in the table, probably best termed a perigynous or epigynous zone, has often been referred to in the botanical literature as a 'floral cup or tube' or 'hypanthium'. These terms have generally been used when sepals, petals and stamens are all inserted on such a ring of tissue, as in *Prunus* or *Rhamnus*.

In flowers with a superior ovary, the stamens may be apparently borne on the corolla (e.g. Primulaceae, Malvaceae, etc.); in these cases it seems reasonable to describe the corolla plus androecium (stamens considered as a whole) as perigynous and the calyx as hypogynous. In such flowers there is no term in general use for the tissue between the insertion of the stamens on the corolla and the base of the united organs; however, there seems no reason why it should not be termed (if a term is needed) a perigynous zone (it is not, of course, possible to consider it a hypanthium). In other flowers the calyx and corolla are apparently fused to each other below, whereas

Table 1. *Relationships of floral parts (see fig. 1)*

Ovary (G) position	Fig. 1	Insertion of perianth (P or K & C) and androecium (A)	Description adopted here	Description used in older literature
	(a)	PA or KCA inserted independently on the torus (e.g. *Ranunculus*)	PA or KCA hypogynous	Flower hypogynous
	(b)	K & C apparently fused at the base, A inserted independently on the torus (e.g. *Tropaeolum*)	K & C perigynous borne on a perigynous zone A hypogynous	Various
Superior	(c)	C & A aparently fused at base, K inserted independently on torus (e.g. *Primula*)	K hypogynous C & A perigynous borne on a perigynous zone	Flower hypogynous A epipetalous
	(d)	K, C & A inserted on a ring or collar of tissue which is inserted on the torus (e.g. *Prunus*)	K, C & A perigynous borne on a perigynous zone	Flower perigynous
	(e)	P & A apparently fused, C absent (e.g. *Daphne*)	P & A perigynous	Various

6

Table 1. *cont.*

Ovary (G) position	Fig. 1	Insertion of perianth (P or K & C) and androecium (A)	Description adopted here	Description used in older literature
Partly inferior	(f)	P & A or K, C & A inserted independently apparently on walls of ovary (e.g. *Paliurus*, some species of *Saxifraga*)	P & A or K, C & A partly epigynous	Various
	(g)	P & A or K, C & A inserted independently apparently on top of the ovary (e.g. Umbelliferae)	P & A or K, C & A epigynous	Flower epigynous
Fully inferior	(h)	K, C & A inserted on top of ovary, C & A fused (e.g. *Viburnum*)	K, C & A epigynous, C & A borne on an epigynous zone	Flower epigynous A epipetalous
	(i)	K, C & A inserted on a ring or collar of tissue itself inserted on top of the ovary (e.g. *Fuchsia*)	K, C & A epigynous, C & A borne on an epigynous zone	Flower epigynous

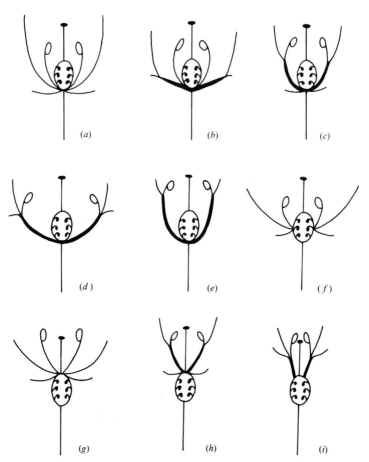

Figure 1. Diagrams illustrating the usage of the terms hypogyny, perigyny and epigyny. Perigynous and epigynous zones are indicated by the use of heavy lines. For further information see table 1.

the stamens are quite free (e.g. *Tropaeolum*). Again, there is no reason why the portion of tissue apparently composed of calyx and corolla should not be called a perigynous zone.

When the ovary is inferior, the stamens may be free from the corolla (e.g. *Vaccinium*) or fused to it (e.g. Solanaceae). In the latter case, the fused tissue composed of corolla plus stamens may be termed an epigynous zone (again, not equivalent to a hypanthium). In *Fuchsia* (Onagraceae), petals, sepals and stamens are all borne on the top of such a zone.

A warning must be given here about unisexual flowers. If a pistillode is present in a male flower, it is possible to decide whether the perianth and stamens are hypogynous, perigynous or epigynous. If, however, no pistillode is present, care must be taken. When the ovary of a female flower is partly or completely inferior, it seems reasonable to treat the perianth and stamens of the male flower as epigynous; when the ovary in the female flower is superior, then, if the perianth and stamens in the male flower are inserted independently on the torus, they are treated as hypogynous; if there is any apparent fusion (i.e. if the stamens are actually borne on the perianth), as perigynous. If the female flower lacks a perianth (e.g. *Betula*), then, of course, it is not possible to use these terms at all; we have used the term 'ovary naked' to refer to this condition.

The following situations occur in the families covered by the key (though other patterns occur in some tropical and southern hemisphere families):

I.   Perianth and stamens hypogynous; ovary superior (e.g. *Ranunculus*, *Geranium*, *Cistus*, *Silene*, etc.). See fig. 2(*a*), (*b*).

II. Sepals hypogynous, petals and stamens perigynous; ovary superior (e.g. most Malvaceae, Primulaceae, Scrophulariaceae, etc.). See fig. 2(*c*), (*d*).

III. Perianth (sepals & petals) perigynous; stamens hypogynous (e.g. *Tropaeolum*, *Aesculus*). See fig. 2(*e*).

IV. Perianth and stamens perigynous, ovary superior (e.g. *Prunus*, *Geum*, *Bergenia*, *Staphylea*, *Daphne*, etc.). See fig. 3(*a*)–(*c*).

V. Ovary partly or fully inferior; perianth and stamens epigynous, without an epigynous zone (e.g. Umbelliferae, *Campanula*, *Vaccinium*, etc.). See fig. 3(*d*), (*e*).

VI. Ovary partly or fully inferior, perianth and stamens borne on an epigynous zone composed of 3 whorls (e.g. *Ribes*, *Fuchsia*, *Leptospermum*) or of 2 whorls only (e.g. *Viburnum*, Compositae, etc.). See fig. 4(*a*), (*b*).

Types III and IV are often complicated by the presence of a disc (usually nectar-secreting) surrounding and sometimes almost covering the ovary. Usually, when such a disc is present, the perianth (as a clearly recognisable structure) is inserted on the edge of it, and may therefore be said to be perigynous. The stamens may be borne on the edge of the disc, as in many species of *Acer* (see fig. 4(*c*)), when they are considered as perigynous, or on the top of the disc, as in many species of *Euonymus* (fig. 4(*d*)) when they are treated as hypogynous. A further complication arises in *Passiflora* (fig. 4(*e*)), in which the stamens and ovary are elevated on a common stalk (androgynophore). In the key, *Passiflora* is treated with those families in which the stamens are hypogynous.

In conclusion, it must be stressed that these features must be looked for in a fully open flower; in later stages the relationships of the parts may change. However, it is usually possible to tell that a fruit has developed from a superior or inferior ovary from the positions of the scars of the fallen floral parts – if the scars surround the base of the fruit then the ovary was superior, whereas if they are found on or near the top of the fruit, then the ovary was probably inferior.

We have followed tradition in not applying the hypogyny, perigyny or epigyny terminology to the Monocotyledons. In the descriptions of the Dicotyledonous families (pp. 73–105) the terms (abbreviated to 'hypog', 'perig' or 'epig') are used; the ovary position is not normally given, but may easily be found by reference to table 1 (p. 6) and fig. 1 (p. 8).

Placentation

Although there is some overlap between the different types of placentation, we may conveniently deal with the character under the following headings, using the terms as they are used in the key. Placentation should ideally be observed in both transverse and longitudinal sections of the ovary.

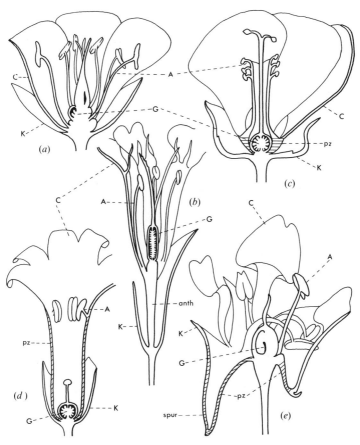

Figure 2. Relative positions of floral parts (see p. 9). Type I: (*a*) *Geranium*, (*b*) *Silene*; type II: (*c*) *Abutilon*, (*d*) *Primula*; type III: (*e*) *Tropaeolum*. A – androecium, anth – anthophore, C – corolla, G – gynoecium, K – calyx, pz – perigynous zone (shaded).

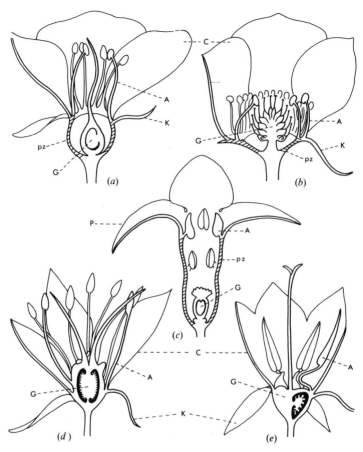

Figure 3. Relative positions of floral parts (see p. 9). Type IV: (*a*) *Prunus*, (*b*) *Geum*, (*c*) *Daphne*; type V: (*d*) *Saxifraga stolonifera*, (*e*) *Campanula*. A – androecium, C – corolla, G – gynoecium, K – calyx, P – perianth (undifferentiated), pz – perigynous zone (shaded).

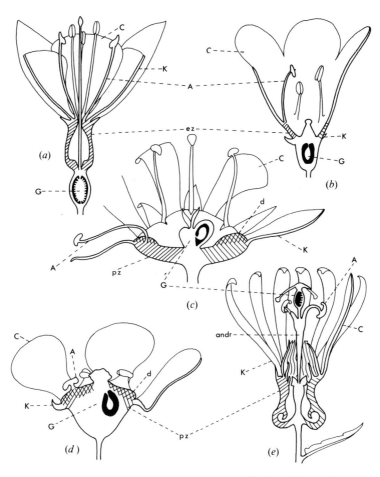

Figure 4. Relative positions of floral parts (see p. 10). Type VI: (*a*) *Fuchsia*, (*b*) *Viburnum*. More complicated types: (*c*) *Acer* (see p. 10), (*d*) *Euonymus* (see p. 10), (*e*) *Passiflora* (see p. 10). A – androecium, andr – androgynophore, C – corolla, d – disc (cross-hatched), ez – epigynous zone (hatched), G – gynoecium, K – calyx, pz – perigynous zone (hatched).

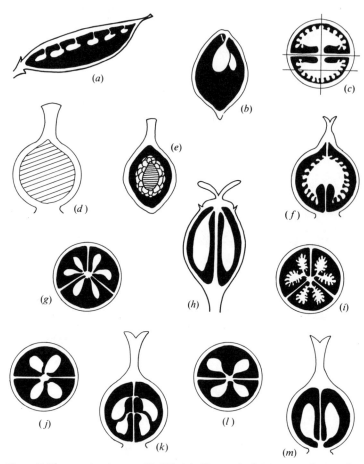

Figure 5. Placentation (see pp. 10–19). (*a*), (*b*) marginal; (*c*)–(*m*) axile. (*c*)–(*f*) ovules on swollen placentas ((*c*) transverse, (*d*)–(*f*) longitudinal sections; planes of the longitudinal sections indicated in (*c*)). (*g*) ovules borne on the axis. (*h*) ovules pendulous. (*i*) ovules on intrusive placentas. (*j*), (*k*) ovules superposed. (*l*), (*m*) ovules side-by-side.

1. *Marginal.* This is used only in those cases in which the carpels are free, and describes the condition in which the carpel bears several ovules on its upper suture (e.g. *Caltha, Pisum*, etc., see fig. 5(*a*), (*b*)).

2. *Axile.* Here the ovary is always made up of fused carpels and contains cross-walls (septa). The ovules are borne on the central axis (e.g. *Narcissus*, see fig. 5(*g*)), on swollen placentae (e.g. *Solanum*, see fig. 5(*c*)–(*f*)) or on intrusive placental outgrowths (e.g. *Begonia*, see fig. 5(*i*)). In some families the ovules are reduced to 1 or 2 in each cell and ascend from the base (e.g. *Ipomoea*, see fig. 5(*l*), (*m*)) or are pendulous from the apex (e.g. Umbelliferae, see fig. 5(*h*)). Ovules in axile ovaries sometimes occur side-by-side (collateral) as in *Heliotropium* (see fig. 5(*l*), (*m*)) or one above the other (superposed) as in Acanthaceae (see fig. 5(*j*), (*k*)). Occasionally axile ovaries are further divided by secondary septa which grow inwards from the carpel wall as the ovary matures (e.g. *Linum, Salvia*), so that the ovary comes to have twice as many cells as carpels.

3. *Parietal.* This term is used when the ovules are borne on the walls of the ovary, or on outgrowths from them. Several situations may be distinguished.

In the majority of cases parietal placentation occurs in 1-celled (unilocular) ovaries made up of several united carpels, the ovules being restricted to placental regions on the walls (usually interpreted as carpel-margins), as in *Viola* (see fig. 6(*a*)), *Gentiana* (see fig. 6(*d*)–(*f*)) or *Ribes*, or on intrusive, placenta-bearing outgrowths from them (e.g. *Cistus, Heuchera*, see fig. 6(*b*)). Intrusive parietal placentas may nearly meet in the middle of the ovary, so that the distinction between axile and parietal placentation is not always clear-cut (e.g. *Escallonia, Cucumis*, see fig. 6(*g*)–(*i*)).

In a few cases, the ovules are borne on the walls of a 2- or more-celled ovary. There may be a development of a secondary, false septum, as in Cruciferae (see fig. 6(*c*)), or the ovary may be initially divided into cells as in most Aizoaceae.

Occasionally the ovules are scattered over most of the carpel surfaces. This situation is distinguished as diffuse-parietal placentation, and can occur in ovaries of free carpels (e.g. Butomaceae, see fig. 6(*j*)) or of united carpels (Hydrocharitaceae, see fig. 6(*k*)).

As shown in the accompanying diagrams, the side view of both axile and parietal placentation can vary greatly according to the

Figure 6. Placentation (see pp. 10–19). Parietal types. (*a*) ovules on the carpel walls. (*b*) ovules on intrusive placentas. (*c*) ovules on the carpel walls, septum present. (*d*)–(*f*) ovules on carpel walls: (*d*) longitudinal section through placentas, (*e*) transverse section, (*f*) longitudinal section at right angles to placentas. (*g*)–(*i*) ovules on intrusive placentas which almost meet in the centre of the ovary; (*g*) longitudinal section through the placentas, (*h*) transverse section, (*i*) longitudinal section between the placentas. (*j*), (*k*) diffuse parietal.

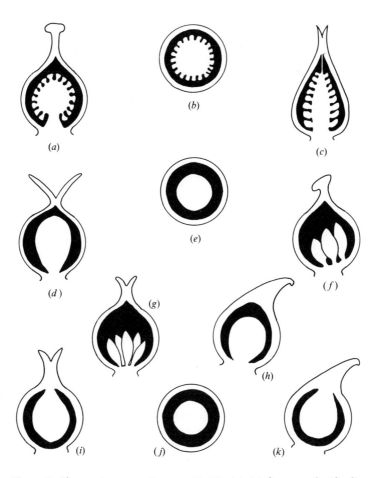

Figure 7. Placentation types (see pp. 10–19). (*a*)–(*c*) free-central; (*d*)–(*h*) basal; (*i*)–(*k*) apical. (*a*), (*b*) ovules free-central. (*c*) ovules free-central showing attachment of placenta to top of ovary. (*d*), (*e*) one basal ovule. (*f*) ovules on an oblique placental cushion. (*g*) several basal ovules (ovary of united carpels). (*h*) one basal ovule (ovary of free carpels). (*i*), (*j*) ovule apical (ovary of united carpels). (*k*) ovule apical (ovary of free carpels).

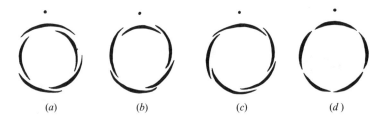

Figure 8. Aestivation types. (*a*), (*b*), overlapping (imbricate) – the details of the manner in which the organs overlap each other is variable and two of the common types are shown. (*c*) contorted (each overlapping one other and overlapped by one other – a special case of imbrication). (*d*) edge-to-edge (valvate).

vertical plane in which the ovary has been cut. It can be best understood in relation to a transverse section.

4. *Free-central*. In this condition the ovules (usually many) are borne on a central spherical or columnar structure that rises from the base of a 1-celled ovary made up of several united carpels (e.g. *Pinguicula*, see fig. 7(*a*)–(*c*)). In most cases a thread of tissue attaches this placental column to the top of the ovary; sometimes this thread is rather stout (e.g. *Lysimachia*, see fig. 7(*c*)). Occasionally the ovary may be septate near the base (e.g. *Silene*) although in most ovaries with free-central placentation the septa break down as the ovary matures.

5. *Basal*. Here the ovules arise from the base of a 1-celled ovary (e.g. *Polygonum*, *Tamarix*, *Armeria*, see fig. 7(*d*), (*e*), (*g*)) or are borne on a basal placental cushion (oblique in *Berberis*, see fig. 7(*f*)). The term can be applied to ovaries made up of several united carpels as well as to individual free carpels (see fig. 7(*h*)).

6. *Apical*. In this case the ovule is attached to the apex of the single cell, as in *Scabiosa* (see fig. 7(*i*), (*j*)) or *Anemone* (see fig. 7(*k*)); here again, the term may be used for an ovary of several united carpels or for an individual free carpel.

Although it would be possible to describe the ovules in the cells of many septate ovaries as apical or basal, we have not used the terms to cover these situations. Instead, to avoid confusion, 'pendulous' or 'ascending' have been used instead, and are considered to be forms of axile placentation.

Aestivation

The relationships of the lobes of the perianth in bud, known as aestivation, form another useful character that sometimes causes difficulty. In the key, 3 basic types are used, illustrated in fig. 8: these are (*a*) and (*b*) parts overlapping in various manners but not as in (*c*) below (imbricate); (*c*) parts overlapping so that each part is overlapped by one and overlaps another part (contorted); (*d*) parts edge-to-edge (valvate). The character is best observed in a transverse section of a flower-bud just before opening, but may be deduced from the positions of the bases of the parts (sepals and petals, or perianth-segments) of a mature flower. One rarely occurring type, which is not illustrated, is seen in the petals of *Papaver* and *Lythrum*, which are irregularly crumpled.

Examining the plant

Using the keys provided in this book, the only tools usually essential for identifying an unknown flowering plant to its family are a good hand-lens and a razor blade. If working indoors, a pair of dissecting needles and a dissecting microscope are helpful.

It is assumed that the user has some familiarity with plant morphology – that he or she will give particular attention to the arrangement of floral parts and to their positions of attachment, and that he will count the numbers of parts when these are not numerous. A flower is usually examined from the outside inwards, the parts being counted before the flower is sectioned. If the flower is bilaterally symmetric (zygomorphic) it must be sectioned along the plane of symmetry.

In the following paragraphs, attention is drawn mainly to features that may cause difficulty when using the key, either because they are readily overlooked or difficult to interpret. The previous section should be consulted for placentation, aestivation and the relative positions of the floral parts. Reasonably complete specimens, preferably including leaves, flowers (of both sexes, where appropriate) and fruits are desirable. In the keys we have tried to use rootstock and fruit characters only when other, generally more readily available vegetative or floral characters are insufficient for certain identification.

Vegetative characters

Leaf position and blade. In plants with deciduous leaves, flowering before the new leaves appear, leaf position (alternate, opposite, whorled) can be deduced by examination of the leaf-scars. In the key, 'opposite' leaves should usually be interpreted as including the whorled condition unless otherwise specified. We have used the term 'divided' to apply to any deep division of a simple (i.e. not strictly compound) leaf. For the use of the term 'herb' see the glossary.

Stipules. Stipules may be present, though often minute, or absent;

normally at the base of the leaf-blade or stalk, they may stand between two opposite leaves (interpetiolar stipules), in which position they are often united, or rarely between the leaf-base and the stem (intrapetiolar stipules); the last two conditions both occur in Rubiaceae. It is important to remember that stipules are often deciduous; they should be looked for on young shoots, or an older shoot should be examined for stipular scars.

Other distinctive features to look for include translucent glands (best seen as clear spots when a leaf is held up to a bright light) which may or may not be aromatic when the leaf is crushed, milky or coloured sap, and distinctive hair types (e.g. stellate or medifixed hairs, lepidote scales).

Floral characters

Except for the way the perianth is arranged in bud (see previous section), floral characters, including placentation, should be observed at full flowering, since this is the stage in which they are recorded in the key.

Observe whether a flower is bisexual (hermaphrodite) or unisexual (and, if so, whether the plant is monoecious or dioecious); in some cases styles and non-functional ovaries (pistillodes) may be present in male flowers (*Ilex*), or non-functional stamens in the female (*Acer*). A radially symmetric (actinomorphic) flower has sometimes only 2 planes of symmetry (Papaveraceae) but usually more. A bilaterally symmetric (zygomorphic) flower has only 1 plane of symmetry. These two terms, denoting type of symmetry, may apply to the whole flower or to parts of it (particularly the corolla). By convention the ovary is not taken into consideration when deciding on the symmetry of a particular flower. A few flowers (e.g. *Canna*) have no plane of symmetry and are termed asymmetric; in some species of *Maranta* the flowers are borne in pairs; each flower is asymmetric but the pair considered together is bilaterally symmetric.

Perianth. It is important to observe whether the perianth is in one or more series, or absent; if in series, whether the members of the various series resemble one another in form, colour and texture. A warning is needed concerning the calyx. This may be very quickly deciduous, being thrown off as the flower opens (e.g. *Papaver*), so

21

that the mature flower appears to have a 1-seriate perianth. In some genera, especially those with condensed inflorescences and/or inferior ovaries, the calyx-teeth are minute, being reduced to a rim, as in many Umbelliferae, Rubiaceae, Compositae and some species of *Rhododendron*). A 1-seriate perianth may sometimes be conspicuous, a tubular or petal-like structure (as in many Thymelaeaceae and Nyctaginaceae) and must not be confused with the corolla of a genus with a reduced calyx. In some cases an epicalyx may be present outside the true calyx (e.g. *Geum*, *Hibiscus*); in others the perianth may bear a petal-like or filamentous outgrowth called a corona (e.g. *Narcissus*, *Passiflora*).

It is sometimes difficult to see if all the petals are united at the base. The basal zone of union may be very short, and can most easily be observed by gently pulling off the whole corolla, or by observing how the corollas fall naturally. Care must be taken with those plants in which the petals are free at the base but united or closely coherent higher up (e.g. some Leguminosae, Stackhousiaceae).

A nectar-secreting disc, usually forming a ring or tube is often present in many insect-pollinated flowers – see p. 10 for a discussion of such discs. In other plants nectar is secreted by separate glands (e.g. *Geranium*), or even by the perianth itself, often in sacs or spurs developed on the corolla (e.g. *Delphinium*). In many Malvaceae it is secreted by nectaries borne on the calyx.

Stamens (Androecium).   In a few families the stamens are equal in number to (very rarely fewer than) the petals and on the same radii as them (i.e. they are antipetalous). This is an almost constant feature in Berberidaceae, Lardizabalaceae, Sabiaceae, Rhamnaceae, Leeaceae, Vitaceae, Myrsinaceae, Primulaceae and Plumbaginaceae. In the condition known as obdiplostemony (e.g. *Geranium*) the stamens of the outer of 2 whorls are on the same radii as the petals (as opposed to the more common situation in which those of the inner whorl are on these radii). In some groups the anthers, instead of opening by the more usual longitudinal slits, open by terminal pores. Such poricidal dehiscence of the anthers is characteristic of the following families: Ochnaceae, Elaeocarpaceae, Tremandraceae, Polygalaceae, Melastomataceae, Actinidiaceae (part), Ericaceae (most), Mayacaceae, etc., but occurs sporadically in other groups (e.g. *Galanthus* in the Amaryllidaceae). Occasionally, the anthers open by flap-like, lateral valves (e.g. Lauracaeae, many Berberidaceae, Monimiaceae and Hamamelidaceae).

Pollen is sometimes shed in tetrads (Juncaceae, Droseraceae, most Ericaceae), but this condition generally requires observation through a compound microscope. Pollen in coherent masses (pollinia) occurs in most Asclepiadaceae and Orchidaceae, as well as in some Leguminosae.

Ovary. The importance of placentation has been stressed above (p. 10). It is best observed by cutting two ovaries, one transversely, the other longitudinally (in the plane of symmetry of the flower if the flower is bilaterally symmetric). When the ovary is very small in flower, it is often helpful to cut one that is beginning to swell and ripen. Ovule attachment may sometimes be seen more easily by slitting the side of the ovary with a needle, rather than by sectioning (e.g. Compositae, Cyperaceae). The number of ovules, though not their attachment, can often be seen by gently squeezing the ovary so that the ovules pop up. It should be noted that the number of ovules is often greatly reduced after fertilisation.

Difficulty is sometimes experienced in deducing the number of carpels united into a compound ovary, or indeed, whether an ovary consists of a single carpel or of several united together. Apart from developmental and anatomical studies, which raise as many problems as they solve, morphological information from various sources helps.

1. The number of styles or stigmas is usually a reliable indicator, and the one that is most widely applicable. As a general rule, the number of stigmas equals the number of carpels. However, divided stylar arms may mislead one into thinking that there are more carpels than there really are (e.g. *Euphorbia*). No help can be derived from stigma number if the stigma is capitate or punctate.

2. The number of placentae in an ovary with axile or parietal placentation corresponds with the number of carpels, although the correlation may be disturbed by placental modifications (e.g. *Orobanche*). In ovaries with free-central, basal or apical placentation, the position of the ovules is no help.

3. A single, complete septum and 2 cells indicates an ovary of 2 carpels (e.g. *Antirrhinum*). When more than 1, the number of septa can be misleading. In some cases (e.g. *Viburnum*, many Valerianaceae) only 1 cell is fertile, the others being reduced.

23

4. When the fruit is a capsule, it generally splits into valves that correspond with the number of carpels. However, some capsules have twice as many valves (teeth) as there are carpels due to splitting of the capsule teeth themselves (e.g. *Silene*). In schizocarpic fruits, the number of mericarps corresponds to carpel number (e.g. Umbelliferae, *Geranium*), unless there has been false septation (e.g. Labiatae).

In the keys and descriptions, when an ovary is described as having a certain number of cells without any mention of the carpel number, this is generally because the latter is uncertain, or the number of cells easier to determine. Indeed, in much modified, 1-celled ovaries (e.g. *Pandanus*, *Mirabilis*, *Berberis*) estimation of carpel number largely depends on the interpretation of vascular anatomy.

Using the keys

The keys used here are of the bracketed type, and are dichotomous throughout, i.e. at every stage a choice must be made between two (and only two) contrasting alternatives (leads), which together make up a couplet. As the main key allows for the identification of 257 families, it has been arranged in groups, with a key to the groups at the beginning. To facilitate reference to particular leads, each couplet is numbered and each lead is given a distinguishing letter (a or b). To find the family to which a specimen belongs, one starts with the key to the groups and compares the specimen with the two leads of the couplet numbered 1. If the plant agrees with 1a, one proceeds to the lead with the number that is the same as that appearing at the right hand end of 1a (in this case, 2); if, however, the plant agrees with 1b, then one proceeds to the couplet numbered 13. This process is repeated for subsequent couplets until, instead of a number at the right hand end of a lead, a group is reached. It is very important that the whole of each couplet is read and understood.

One proceeds in the same way within the group keys until the name of a family is reached. The families are numbered in the key, and, to provide a check on the identification obtained, the families are briefly described, in numerical order on pp. 73–114. The specimen should be compared carefully with the description of the family; this should help to reveal errors in identification and observation.

In order that back-checking should be easy, the number of the lead from which any particular couplet is derived is given in brackets after the couplet number of the 'a' lead. Thus '13a. (3)' means that one lead of couplet 3 takes one directly to couplet 13.

It will sometimes happen that the specimen does not agree with all the characters given in a particular lead. When this situation arises, one must decide which of the two leads of the relevant couplet the specimen agrees with most fully. In general, we have put the most reliable diagnostic characters at the beginnings of the leads, so these characters should be observed with particular care. The only exception to this procedure occurs when the second lead reads 'Combination of characters not as above'. In such cases the specimen must agree with all of the characters given in the 'a' lead; if it deviates in one or more characters from those detailed in the 'a' lead, it must be treated as falling into the 'combination of characters not as above' category.

Experience with earlier editions of this key has shown that users go wrong more often with the key to the groups than elsewhere; and, of course, an error here means that a correct identification is virtually impossible. The following paragraphs consist of a commentary on the key to the groups, with precise indications of how it should be used. The principles covered by this commentary, if not the details, are also relevant to the rest of the key.

Couplet 1 discriminates between the Dicotyledones and the Mono-cotyledones, the two large groups into which the Flowering Plants (Angiospermae) naturally divide. There is no single character that completely and certainly distinguishes these two groups; instead, a combination of a rather large number of characters has to be used, and couplet 1 includes the most readily observed of these. 1a. starts with the phrase 'Cotyledons usually 2, lateral', as opposed to 1b. 'Cotyledon 1, terminal'. Specimens with 2 cotyledons (seedling leaves) clearly belong to 1a, but those with only 1 cotyledon could fit either, though it must be borne in mind that among the Dicotyledones the occurrence of plants with only 1 cotyledon is extremely rare. Of course, the chances of the user being able to answer this question if he has only a mature specimen are very small; however, the character is important enough to deserve mention. The second phrases of the couplet are: 'leaves usually net-veined, with or without stipules, alternate, opposite or whorled' as opposed to 'leaves usually with parallel veins sometimes connected by cross-veinlets; leaves without stipules, opposite only in some aquatic plants'. These are admittedly very imprecise alternatives, but they do help in distinguishing the two groups. Firstly, if the specimen has stipules (or scars left by them), or if it is a terrestrial plant with opposite leaves, then clearly it belongs to 1a. If the leaves are net-veined, then there is a high probability that it belongs to 1a, just as, if the leaves have parallel veins, there is a high probability that it belongs to 1b.

The third characteristic used in this couplet reads: 'flowers with parts in 2s, 4s, 5s or parts numerous' as opposed to 'flowers usually with parts in 3s'. 'Parts', here, essentially means sepals and petals or perianth and stamens. Again, this is a matter of probabilities: if the flower has parts in 2s, 4s, 5s (or multiples of these) or the parts are numerous, then there is a high probability that it belongs with 1a; if the parts are in 3s (or multiples) then there is a high probability that it belongs with 1b. Finally, the phrase 'primary root system (taproot) usually persistent, branched' is opposed to 'mature root system wholly adventitious' again poses a similar set of probabilities.

In making a decision about a particular plant, it is necessary to observe what information is available and to make a judgement as to the balance of probabilities. Fortunately, with experience, the discrimination of these two large groups is actually easier than it appears, and little difficulty is generally found in deciding whether or not a particular specimen fits one or the other.

Couplet 2 (arrived at from 1a) is much more straightforward. If the plant has petals which are united to each other into a cup or tube at the base, then the plant matches 2b. If, on the other hand, the petals are quite separate from each other at the base, or if the flower has no petals at all, then the plant matches 2a. Some problems may arise in deciding whether or not there are any petals present: these problems also occur with couplet 5, and are discussed below.

Couplet 3 (arrived at from 2a) separates off what are essentially the catkin-bearing trees and shrubs. Plants matching 3a are always woody and have unisexual flowers with at least the males in catkins. A catkin is an inflorescence of small flowers without petals which are borne in the axils of bracts which overlap and protect the flowers; the anthers (and the stigmas if the female flowers are also in catkins) project between the bracts when they are ripe (or receptive). If these are present then the plant matches 3a; if not, then 3b should be chosen.

Couplet 4 concerns only the ovary. If it is made up of 2 or more separate carpels then the specimen matches 4a. If, on the other hand, the carpels are joined to each other (even if only by the styles), then the specimen matches 4b. If the flower contains only a single carpel (which will have a single style), then it matches 4b.

Couplet 5 deals essentially with the question of whether the perianth is single or double (i.e. consisting of a calyx and corolla). This can be surprisingly difficult to decide in some cases, but the following guidelines may help (see p. 21):

(a) Generally, sepals are greenish, have a single main vein, and enclose the flower in bud; petals are usually coloured, rather flimsy, generally have several main veins, and have no protective function. If both sepals and petals are present, and are of the same number, then the radii on which they stand will generally alternate.

(b) Exceptions to these conditions arise from various causes: (i) The sepals may be very small, playing no part in bud-protection; they can usually be seen as very small points or outgrowths, but must be looked for carefully. (ii) The petals may be quite small and of unusual form, as in many Ranunculaceae (e.g. *Nigella*, *Helleborus*, *Delphinium*, etc.). (iii) The sepals may be coloured and petal-like, as in *Aquilegia*; when this is the case they are usually distinguishable from the petals by their shape and size.

If the petals are genuinely absent, the perianth-segments may still look corolla-like, as in *Anemone*. Such cases are distinguishable from those mentioned under (b)(i) by the complete absence of anything that can be interpreted as a reduced sepal.

In a few cases, especially in the Berberidaceae, the sepals and petals are very similar to each other, differing only in size.

Having studied carefully the perianth of the plant under consideration, and bearing in mind what has been said above, a decision can be made as to whether 5a or 5b is matched. The note about aquatic plants mentioned under 5a must also be considered: such plants are definitely excluded from 5a.

Couplet 6 is relatively simple. If there are more than twice as many stamens as petals then 6a is chosen; if the stamens are fewer than this, then 6b.

Couplet 7 is also simple, requesting a decision as to whether the ovary is superior or inferior. This question is dealt with in detail on pp. 4–5.

Couplet 8 deals with placentation, dealt with on pp. 10–19.

Couplet 9 is the equivalent of couplet 7, but is worded differently to allow for the identification of unisexual flowers which occur with some frequency in the groups deriving from couplet 9.

Couplet 10 deals with the sexuality of the flowers. This is often clear-cut (e.g. in *Begonia*) but is sometimes not so obvious. Care must be taken in the examination of stamens and ovaries as in some functionally unisexual flowers, stamens or ovaries may be present but sterile. Sterile stamens are often smaller than fertile ones (which may be found in other flowers on the same or adjacent plants), with translucent anthers which produce no pollen. Similarly, sterile ovaries are also smaller than the fertile, and contain no plump, healthy-looking ovules.

Couplet 11 covers the same ground as couplets 7 and 9.

Couplet 12 is one that can lead to problems. It asks about the symmetry of the corolla, which seems to be a difficult matter for many people, though it is generally easily demonstrated in particular examples. A radially symmetric (actinomorphic) flower is one which has many planes of symmetry. This is best visualised by making a number of imaginary cuts down several different diameters of the flower (such flowers are generally circular in outline) and noting whether in each case the 2 imaginary halves of the corolla are mirror-images of each other. If this is the case, then the corolla is radially symmetric. If, on the other hand, the corolla appears compressed (usually from side to side) and only 1 imaginary cut produces mirror-image halves (generally this cut is in the vertical plane), then the corolla is bilaterally symmetric (zygomorphic). It is important to notice in this couplet that only the condition of the corolla is asked for. In flowers with radially symmetric corollas the stamens can be irregularly arranged (e.g. deflected downwards) rendering the flower considered as a whole as bilaterally symmetric, though the corolla itself is radially symmetric.

Couplet 13 deals with the Monocotyledones, breaking them down into two large groups on the basis of ovary position. The qualifying phrases should be noted, in that 13a includes all submerged aquatic Monocotyledones.

Proceeding through the keys in this manner, making careful observations of the plant in the light of what is understood from a detailed reading of the couplets, should lead to accurate identifications. One further couplet that has given trouble in the past is Group B 4. Though somewhat complex, this couplet separates quite a few families, and is worth proper understanding. Lead 4a includes herbs, succulent shrubs, shrubs with yellow wood and climbers with bisexual flowers together with opposite leaves. Lead 4b covers trees and shrubs which are neither succulent, nor with yellow wood, nor climbers with both bisexual flowers and opposite leaves (i.e. it does cover climbers with bisexual flowers and alternate leaves, or climbers with unisexual flowers). The contrast is quite clear when understood.

Keys

Key to Groups

1a. Cotyledons usually 2, lateral; leaves usually net-veined, with or without stipules, alternate, opposite or whorled; flowers with parts in 2s, 4s or 5s or parts numerous; primary root-system (taproot) usually persistent, branched (*Dicotyledons*)   2

  b. Cotyledon 1, terminal; leaves usually with parallel veins, sometimes these connected by cross-veinlets; leaves without stipules, opposite only in some aquatic plants; flowers usually with parts in 3s; mature root-system wholly adventitious (*Monocotyledons*)   13

2a. (1) Petals present, free from each other at their bases (rarely united above the base), usually falling as individual petals, or petals absent   3

  b. Petals present, all united into a longer or shorter tube at the base, falling as a complete corolla   11

3a. (2) Flowers unisexual and without petals, at least the males borne in catkins which are usually deciduous; plants always woody   **Group A** (p. 32)

  b. Flowers with or without petals, unisexual or bisexual, never in catkins; plants woody or not   4

4a. (3) Ovary consisting of 2 or more carpels which are completely free from each other   **Group B** (p. 33)

  b. Ovary consisting of a single carpel or of 2 or more carpels which are united to each other wholly or in part (rarely the bodies of the carpels free but the style single)   5

5a. (4) Perianth of 2 or more whorls, more or less clearly differentiated into calyx and corolla (calyx rarely very small and obscure; excluding aquatic plants with minute, quickly deciduous petals and branch-parasites with opposite, leathery leaves)   6

  b. Perianth of a single whorl (which may be corolla-like) or perianth completely absent, more rarely the perianth of 2 or

more whorls but the segments not differing from whorl to
whorl   9
6a. (5) Stamens more than twice as many as the petals
                                                    Group C (p. 36)
  b. Stamens twice as many as the petals or fewer   7
7a. (6) Ovary partly or fully inferior   **Group D** (p. 39)
  b. Ovary completely superior   8
8a. (7) Placentation axile, apical, basal or free-central
                                                    Group E (p. 41)
  b. Placentation parietal   **Group F** (p. 46)
9a. (5) Stamens borne on the perianth or ovary inferior (perianth of
         female flowers sometimes very small) **Group G** (p. 48)
  b. Stamens free from the perianth; ovary superior or naked (i.e.
         not surrounded by a perianth)   10
10a. (9) Flowers unisexual   **Group H** (p. 50)
  b. Flowers bisexual   **Group I** (p. 52)
11a. (2) Ovary partly or fully inferior   **Group J** (p. 54)
  b. Ovary completely superior   12
12a. (11) Corolla radially symmetric   **Group K** (p. 55)
  b. Corolla bilaterally symmetric   **Group L** (p. 60)
13a. (1) Ovary superior or flowers completely without perianth
         (including all aquatics with totally submerged flowers)
                                                    Group M (p. 63)
  b. Ovary inferior or partly so (if plant aquatic then flowers borne
         well above the water-level)   **Group N** (p. 67)

GROUP A

1a. Stems jointed; leaves reduced to whorls of scales
                                                    1. Casuarinaceae
  b. Stems not jointed; leaves not as above   2
2a. (1) Leaves pinnate   3
  b. Leaves simple and entire, toothed or lobed (sometimes deeply
         so)   4
3a. (2) Leaves without stipules; fruit a nut   **3. Juglandaceae**
  b. Leaves with stipules; fruit a legume   **80. Leguminosae**
4a. (2) Male flowers with a perianth of 2 segments and 4–5 fertile
         stamens plus 4–5 staminodes; female flowers without a peri-
         anth; shrubs; fruit a syncarp of berries   **71. Bataceae**

4b. Combination of characters not as above 5

5a. (4) Leaves opposite, evergreen, entire; fruit berry-like
162. Garryaceae

b. Leaves alternate, deciduous or evergreen; fruit not berry-like 6

6a. (5) Ovules many, parietal; seeds many, cottony-hairy; male catkins erect with the stamens projecting between the bracts or hanging and with fringed bracts **5. Salicaceae**

b. Ovules solitary or few, not parietal; seeds few, not cottony-hairy; male catkins not as above 7

7a. (6) Leaves dotted with aromatic glands **2. Myricaceae**

b. Leaves not dotted with aromatic glands 8

8a. (7) Styles 3, each often branched; fruit splitting into 3 mericarps: seeds with appendages **94. Euphorbiaceae**

b. Styles 1–6, not branched; fruit and seeds not as above 9

9a. (8) Plants with milky sap **10. Moraceae**

b. Plants with clear sap 10

10a. (9) Male catkins compound, i.e. each bract with 2–3 flowers attached to it; styles 2 **6. Betulaceae**

b. Male catkins simple, i.e. each bract with a single flower attached to it; styles 1 or 3–6 11

11a. (10) Ovary inferior; fruit a nut surrounded or enclosed by a scaly cupule; stipules deciduous; styles 3–6 **7. Fagaceae**

b. Ovary superior; fruit a leathery drupe, cupule absent; stipules absent; style 1 **4. Leitneriaceae**

GROUP B

1a. Trees with bark peeling off in plates, palmately lobed leaves and unisexual flowers in hanging, spherical heads
72. Platanaceae

b. Combination of characters not as above 2

2a. (1) Perianth-segments and stamens borne independently below the ovary, or perianth absent 3

b. Perianth-segments and stamens borne on a rim or cup which is borne below the ovary 23

3a. (2) Aquatic plants with peltate leaves and flowers with 3 sepals
45. Nymphaeaceae

b. Terrestrial plants, or, if aquatic then without peltate leaves and flowers with more than 3 sepals 4

4a. (3) Herbs, succulent shrubs or shrubs with yellow wood, or climbers with bisexual flowers and opposite leaves   5

  b. Trees or shrubs which are neither succulent nor with yellow wood, if climbers then with unisexual flowers and alternate leaves   10

5a. (4) Perianth absent   **47. Saururaceae**

  b. Perianth present   6

6a. (5) Leaves succulent; stamens in 1 or 2 whorls

        74. Crassulaceae

  b. Leaves not succulent; stamens spirally arranged, not obviously in whorls   7

7a. (6) Petals fringed; fruits formed from each carpel borne on a common stalk (gynophore)   **69. Resedaceae**

  b. Petals (when present) not fringed, but sometimes modified for nectar-secretion; fruits formed from each carpel not borne on a common stalk   8

8a. (7) Leaves opposite or whorled; flowers small, stalkless, in axillary clusters; ovule 1, placentation basal

        17. Phytolaccaceae

  b. Combination of characters not as above   9

9a. (8) Sepals differing among themselves, green; stamens ripening from the inside of the flower outwards, borne on a nectar-secreting disc   **53. Paeoniaceae**

  b. Sepals all similar, green or petal-like; stamens ripening from the outside of the flower inwards; nectar-secreting disc absent

        41. Ranunculaceae

10a. (4) Leaves simple   11

  b. Leaves compound   21

11a. (10) Sepals and petals 5 each   12

  b. Sepals and petals not 5 each   14

12a. (11) Leaves opposite; stamens 5–10   **104. Coriariaceae**

  b. Leaves alternate; stamens more than 10   13

13a. (12) Anthers opening by pores   **57. Ochnaceae**

  b. Anthers opening by slits   **52. Dilleniaceae**

14a. (11) Unisexual climbers   15

  b. Erect trees or shrubs, flowers usually bisexual   16

15a. (14) Carpels many; seeds not U-shaped   **32. Schisandraceae**

  b. Carpels 3 or 6; seeds usually U-shaped   **44. Menispermaceae**

16a. (14) Stamens with a truncate connective which overtops the anthers; fruit usually fleshy, formed from the closely continguous products of several carpels (a syncarp); endosperm convoluted   **29. Annonaceae**

   b. Connective of stamens not as above; fruit not as above; endosperm not convoluted   17

17a. (16) Carpels spirally arranged on an elongate receptacle; stipules large, united, early deciduous, leaving a ring-like scar

                                                27. Magnoliaceae

   b. Carpels in 1 whorl; stipules absent, minute or united to the leaf-stalk, not leaving a ring-like scar when fallen   18

18a. (17) Petals present   19

   b. Petals absent   20

19a. (18) Sepals free, overlapping, more than 6; ovule solitary in each carpel   **33. Illiciaceae**

   b. Sepals 2–6, united or if free then edge-to-edge in bud; ovules more than 1 in each carpel   **28. Winteraceae**

20a. (18) Leaves in whorls; flowers bisexual; sepals minute or absent   **39. Eupteleaceae**

   b. Leaves opposite or alternate; flowers unisexual; sepals 4

                                                40. Cercidiphyllaceae

21a. (10) Unisexual climbers or erect shrubs with blue fruits; perianth parts in 3s   **43. Lardizabalaceae**

   b. Erect shrubs, fruits not blue; perianth parts not in 3s   22

22a. (21) Flowers showy, bisexual; leaves not aromatic

                                                53. Paeoniaceae

   b. Flowers inconspicuous, unisexual; leaves aromatic

                                                96. Rutaceae

23a. (2) Leaves modified into insect-digesting pitchers

                                                75. Cephalotaceae

   b. Leaves not modified into pitchers   24

24a. (23) Flowers unisexual; leaves evergreen   **34. Monimiaceae**

   b. Flowers bisexual; leaves usually deciduous   25

25a. (24) Inner stamens sterile; perianth of many segments; leaves usually opposite   **35. Calycanthaceae**

   b. Stamens all fertile; perianth of 4–9 segments; leaves usually alternate   26

26a. (25) Leaves without stipules, entire; flowers solitary and terminal; seed with a divided aril   **54. Crossosomataceae**
  b. Leaves usually with stipules and toothed (if entire then flowers
clustered); seed without an aril   **82. Rosaceae**

GROUP C

  1a. Perianth and stamens borne independently below the superior
ovary   2
  b. Perianth and stamens either borne on the edge of a rim or cup
which is itself borne below the superior ovary, or borne on the
top or the sides of the (partly or fully) inferior ovary   30
  2a. (1) Placentation axile or free-central   3
  b. Placentation marginal or parietal   20
  3a. (2) Plancentation free-central; sepals usually 2
                                             20. Portulacaceae
  b. Placentation axile; sepals more than 2   4
  4a. (3) Leaves all basal, tubular, forming insect-digesting pitchers;
style peltately dilated   **63. Sarraceniaceae**
  b. Leaves not as above; style not peltately dilated   5
  5a. (4) Leaves alternate   6
  b. Leaves opposite   17
  6a. (5) Anthers opening by terminal pores   7
  b. Anthers opening by longitudinal slits   9
  7a. (6) Shrubs with simple leaves without stipules, often covered
with stellate hairs; stamens inflexed in bud; fruit a berry
                                             56. Actinidiaceae
  b. Combination of characters not as above   8
  8a. (7) Ovary deeply lobed, borne on an enlarged receptacle or
gynophore; petals not fringed   **57. Ochnaceae**
  b. Ovary not lobed, not borne as above; petals often fringed
                                             123. Elaeocarpaceae
  9a. (6) Inner whorl of perianth-segments tubular or bifid, nectar-
secreting; fruit a group of partly to fully coalescent follicles
                                             41. Ranunculaceae
  b. Combination of characters not as above   10
10a. (9) Leaves with translucent aromatic glands   **96. Rutaceae**
  b. Leaves without such glands   11

11a. (10) Large tropical trees; sepals 5, all or 2–3 of them enlarged and wing-like in fruit; carpels 3   **58. Dipterocarpaceae**

   b. Combination of characters not as above   12

12a. (11) Stipules absent; leaves evergreen   **59. Theaceae**

   b. Stipules present; leaves usually deciduous   13

13a. (12) Filaments free; anthers 2-celled   14

   b. Filaments united into a tube, at least around the ovary, often also around the styles; anthers often 1-celled   15

14a. (13) Nectar-secreting disc absent; stamens more than 15; leaves simple   **124. Tiliaceae**

   b. Nectar-secreting disc present, conspicuous; stamens 15; leaves dissected   **91. Zygophyllaceae**

15a. (13) Styles divided, several; stipules often persistent; carpels 5 or more   **125. Malvaceae**

   b. Style 1, capitate or lobed, stigmas 1–several; stipules usually deciduous; carpels 2–5   16

16a. (15) Stamens in 2 whorls, those of the outer whorl usually sterile   **127. Sterculiaceae**

   b. Stamens in several whorls, all fertile   **126. Bombacaceae**

17a. (5) Sepals united, falling as a unit; fruit separating into boat-shaped units   **55. Eucryphiaceae**

   b. Sepals and fruit not as above   18

18a. (17) Small trees; stamens with brightly coloured filaments which are at least twice as long as the petals, the anthers forming a circle   **60. Caryocaraceae**

   b. Combination of characters not as above   19

19a. (18) Leaves simple, without stipules, often with translucent glands; stamens often united in bundles   **62. Guttiferae**

   b. Leaves pinnate, without translucent glands; stamens not united in bundles   **91. Zygophyllaceae**

20a. (2) Aquatic plants with cordate leaves; style and stigmas forming a disc on top of the ovary   **45. Nymphaeaceae**

   b. Combination of characters not as above   21

21a. (20) Carpel 1 with marginal placentation   22

   b. Carpels 2 or more, placentation parietal   23

22a. (21) Leaves bipinnate or modified into phyllodes, with stipules   **84. Leguminosae**

22b. Leaves various but not as above, without stipules

41. Ranunculaceae

23a. (21) Leaves opposite   24

  b. Leaves alternate   26

24a. (23) Styles numerous; floral parts in 3s   **66. Papaveraceae**

  b. Styles 1–5; floral parts in 4s or 5s   25

25a. (24) Style 1; stamens not united in bundles; leaves without translucent glands   **135. Cistaceae**

  b. Styles 3–5, free or variously united; stamens united in bundles (sometimes apparently free); leaves with translucent or blackish glands   **62. Guttiferae**

26a. (23) Small trees with aromatic bark; filaments of the stamens united   **31. Canellaceae**

  b. Herbs, shrubs or trees, bark not aromatic; filaments free   27

27a. (26) Trees; leaves with stipules; anthers opening by short, pore-like slits   **136. Bixaceae**

  b. Herbs or shrubs; leaves usually without stipules; anthers opening by longitudinal slits   28

28a. (27) Sepals 2 or rarely 3, quickly deciduous **66. Papaveraceae**

  b. Sepals 4–8, persistent in flower   29

29a. (28) Ovary closed at apex, borne on a stalk (gynophore); none of the petals fringed   **67. Capparaceae**

  b. Ovary open at apex, not borne on a stalk; at least some of the petals fringed   **69. Resedaceae**

30a. (1) Flowers unisexual; leaf-base oblique   **143. Begoniaceae**

  b. Flowers bisexual; leaf-base not oblique   31

31a. (30) Placentation free-central; ovary partly inferior

20. Portulacaceae

  b. Placentation not free-central; ovary either completely superior or completely inferior   32

32a. (31) Aquatic plants with cordate leaves   **45. Nymphaeaceae**

  b. Terrestrial plants; leaves various   33

33a. (32) Carpels 1 or 3, eccentrically placed at the top of, the bottom of, or within the tubular perigynous zone

83. Chrysobalanaceae

  b. Carpels and perigynous zone not as above   34

34a. (33) Stamens united into bundles on the same radii as the

petals; staminodes often present; plants usually rough with
stinging hairs   **141. Loasaceae**
  b. Combination of characters not as above   35
35a. (34) Sepals 2, united, falling as a unit as the flower opens;
plants herbaceous   **66. Papaveraceae**
  b. Sepals 4–5, usually free, not falling as a unit as the flower opens;
trees or shrubs   36
36a. (35) Stamens united into several rings or sheets
149. Lecythidaceae
  b. Stamens not as above   37
37a. (36) Carpels 8–12, superposed   **148. Punicaceae**
  b. Carpels fewer, side by side   38
38a. (37) Leaves with stipules   39
  b. Leaves without stipules   40
39a. (38) Leaves opposite or in whorls; plants woody
77. Cunoniaceae
  b. Leaves alternate; plants woody or herbaceous   **82. Rosaceae**
40a. (38) Leaves with translucent aromatic glands; style 1
147. Myrtaceae
  b. Leaves without such glands; styles usually more than 1
76. Saxifragaceae

GROUP D

  1a. Petals and stamens numerous; plant succulent   2
  b. Petals and stamens each fewer than 10; plants usually not
succulent   3
  2a. (1) Stems succulent, often very spiny; leaves usually absent,
very reduced or falling early   **26. Cactaceae**
  b. Stems and leaves succulent, spines usually absent
19. Aizoaceae
  3a. (1) Anthers opening by terminal pores   4
  b. Anthers opening by longitudinal slits   5
  4a. (3) Filaments each with a knee-like joint below the anther;
leaves with 3 conspicuous main veins from the base
150. Melastomataceae
  b. Filaments straight; leaves with a single main vein
151. Rhizophoraceae

5a. (3) Placentation parietal, placentas sometimes intrusive   6

5b. Placentation axile, basal, apical or free-central   7

6a. (5) Climbing herbs with tendrils; flowers unisexual
    144. Cucurbitaceae

  b. Erect herbs or shrubs, if climbing then without tendrils; flowers usually bisexual   **76. Saxifragaceae**

7a. (5) Stamens on the same radii as the petals; trees or shrubs with simple leaves   **120. Rhamnaceae**

  b. Stamens not on the same radii as the petals or more numerous than them; plants herbaceous or woody, leaves simple or compound   8

8a. (7) Flowers borne in umbels, sometimes condensed into heads or superposed whorls; leaves usually compound   9

  b. Flowers not in umbels; leaves usually simple   10

9a. (8) Fruit splitting into 2 mericarps; flowers usually bisexual; petals overlapping in bud; usually herbs without stellate hairs
    164. Umbelliferae

  b. Fruit a berry; flowers often unisexual; petals edge-to-edge in bud; plants mostly woody, often with stellate hairs
    163. Araliaceae

10a. (8) Style 1   11

  b. Styles more than 1, often 2 and divergent   19

11a. (10) Floating aquatic herb; leaf-stalks inflated **146. Trapaceae**

  b. Terrestrial herbs, trees or shrubs; leaf-stalks not inflated   12

12a. (11) Small, low shrubs with scale-like, overlapping leaves; flowers in heads   **81. Bruniaceae**

  b. Trees, shrubs or herbs with expanded leaves; flowers not usually in heads   13

13a. (12) Ovary 1-celled with 2–5 ovules; fruit leathery or drupe-like, 1-seeded   **152. Combretaceae**

  b. Ovary usually with 2–5 cells, ovules various; fruit not as above   14

14a. (13) Ovule 1 in each cell of the ovary   15

  b. Ovules 2–numerous in each cell of the ovary   18

15a. (14) Petals edge-to-edge in bud; flowers usually bisexual   16

  b. Petals overlapping in bud; flowers often unisexual   17

16a. (15) Stamens with swollen, hairy filaments; petals recurved
    158. Alangiaceae

16b. Stamens without swollen, hairy filaments; petals not recurved
161. Cornaceae
17a. (15) Flowers in heads subtended by 2 conspicuous, white
bracts; ovary 6–10-celled **160. Davidiaceae**
b. Flowers various, not as above; ovary 1-celled **159. Nyssaceae**
18a. (14) Sap milky; petals 5; ovary 3-celled
215. Campandulaceae
b. Sap watery; petals 2 or 4; ovary usually 4-celled
153. Onagraceae
19a. (10) Trees or shrubs; hairs often stellate; fruit a few-seeded,
woody capsule **73. Hamamelidaceae**
b. Herbs or shrubs; hairs simple or absent; fruit various, not a
woody capsule **76. Saxifragaceae**

GROUP E

1. Perianth bilaterally symmetric 2
b. Perianth radially symmetric (the stamens sometimes not radially
symmetric due to deflexion) 14
2a. (1) Anthers cohering above the ovary like a cap
111. Balsaminaceae
b. Anthers free, not as above 3
3a. (2) Anthers opening by terminal pores 4
b. Anthers opening by longitudinal slits or by flaps 5
4a. (3) Stamens 8, filaments united for at least half their length;
fruit without barbed bristles **103. Polygalaceae**
b. Stamens 3 or 4, filaments free; fruit covered by barbed bristles
85. Krameriaceae
5a. (3) Plants herbaceous 6
b. Plants woody (trees, shrubs or climbers) 10
6a. (5) Leaves with stipules 7
b. Leaves without stipules 8
7a. (6) Carpel 1; fruit a legume, sometimes 1-seeded
84. Leguminosae
b. Carpels 5; fruit a capsule or berry, or splitting into mericarps
89. Geraniaceae
8a. (6) Sepals, petals and stamens borne independently below the
ovary (rarely the petals and stamens somewhat united at the
base); leaves peltate **90. Tropaeolaceae**

b. Sepals, petals and stamens borne on a rim, cup or tube which is borne below the ovary; leaves not peltate 9

9a. (8) Leaves opposite **145. Lythraceae**

b. Leaves alternate or all basal **76. Saxifragaceae**

10a. (5) Stamens as many as or fewer than petals, borne on the same radii as them **109. Sabiaceae**

b. Stamens more numerous than the petals, or if as many or fewer, not on the same radii as them 11

11a. (10) Carpel 1 with its style arising from near the base

83. Chrysobalanaceae

b. Carpels 2 or more, styles not as above 12

12a. (11) Leaves opposite, palmate; sepals united at the base

108. Hippocastanaceae

b. Leaves alternate, usually pinnate; sepals free 13

13a. (12) Stipules large, borne between the bases of the leaf-stalks

110. Melianthaceae

b. Stipules absent, or, if present, not borne as above

107. Sapindaceae

14a. (1) Anthers opening by terminal pores 15

b. Anthers opening by longitudinal slits or by flaps 22

15a. (14) Leaves with 3 more or less parallel veins from the base; each filament with a knee-like joint below the anther

150. Melastomataceae

b. Leaves with 1 main vein; filaments not as above 16

16a. (15) Low shrubs; leaves and stems covered in conspicuous stalked glandular hairs on which insects are often caught 17

b. Shrubs or rarely low shrubs, not glandular-hairy as above 18

17a. (16) Carpels 2 **79. Byblidaceae**

b. Carpels 3 **80. Roridulaceae**

18a. (16) Low shrubs with unisexual flowers; stamens 4; petals 4, each usually 2–3-lobed, rarely a few unlobed

123. Elaeocarpaceae

b. Combination of characters not as above 19

19a. (18) Carpels 2; leaves opposite **102. Tremandraceae**

b. Carpels 3 or more; leaves alternate 20

20a. (19) Ovary lobed, consisting of several rounded humps, the style arising from the depression between them **57. Ochnaceae**

b. Ovary not lobed as above, style terminal 21

21a. (20) Carpels 3; style divided above into 3 branches; nectar-secreting disc absent **166. Clethraceae**

21b. Carpels usually 4–5, very rarely 3; style not divided; nectar-secreting disc usually present **168. Ericaceae**

22a. (14) Placentation free-central (ovary sometimes with septa below) or basal 23

   b. Placentation axile or apical 27

23a. (22) Stamens as many as petals and on the same radii as them 24

   b. Stamens more or fewer than petals, if as many then not on the same radii as them 26

24a. (23) Anthers opening by flaps; stigma 1 **42. Berberidaceae**

   b. Anthers opening by longitudinal slits; stigmas more than 1 25

25a. (24) Sepals 5; ovule 1, basal on a long, curved stalk; stipules absent **174. Plumbaginaceae**

   b. Sepals 2 or rarely 3; ovules usually numerous, rarely 1 then not on a long, curved stalk; stipules usually present

20. Portulacaceae

26a. (23) Ovary lobed, consisting of several rounded humps, the style arising from the depression between them; leaves pinnatisect **87. Limnanthaceae**

   b. Ovary not lobed, style terminal; leaves simple, entire

22. Caryophyllaceae

27a. (22) Petals and stamens both numerous; plants with succulent leaves and stems **19. Aizoaceae**

   b. Combination of characters not as above 28

28a. (27) Small, hairless annual herb growing in water or on wet mud; seeds pitted **139. Elatinaceae**

   b. Combination of characters not as above 29

29a. (28) Sepals, petals and stamens borne on a rim, cup or tube which is inserted below the ovary 30

   b. Sepals, petals and stamens inserted individually below the ovary 35

30a. (29) Stamens as many as the petals and borne on the same radii as them **120. Rhamnaceae**

   b. Stamens more or fewer than the petals, or if of the same number then not borne on the same radii as them 31

31a. (30) Style 1 32

   b. Styles more than 1, often 2 and divergent 33

32a. (31) Calyx-tube not prominently ribbed; seeds with arils; mostly trees, shrubs or climbers **115. Celastraceae**

b. Calyx-tube prominently ribbed; seeds without arils; mostly herbs **145. Lythraceae**

33a. (31) Fruit an inflated, membranous capsule; leaves mostly opposite, compound **116. Staphyleaceae**

b. Combination of characters not as above 34

34a. (33) Trees or shrubs; hairs often stellate; anthers often opening by flaps; fruit a few-seeded, woody capsule

73. Hamamelidaceae

b. Herbs or shrubs; hairs simple or absent; anthers opening by longitudinal slits; fruit a capsule, not woody **76. Saxifragaceae**

35a. (29) Leaves with translucent, aromatic glands **96. Rutaceae**

b. Leaves without such glands 36

36a. (35) Flowers with a well-developed disc, usually nectar-secreting, below and/or around the ovary 37

b. Flowers without a disc, nectar secreted in other ways 47

37a. (36) Stamens as many as and on the same radii as petals 38

b. Stamens more or fewer than petals, if as many then not on the same radii as them 39

38a. (37) Climbers with tendrils; stamens free **121. Vitaceae**

b. Erect shrubs without tendrils; stamens with their filaments united, at least at the base **122. Leeaceae**

39a. (37) Resinous trees or shrubs 40

b. Herbs, shrubs or trees, not resinous, sometimes aromatic 41

40a. (39) Ovules 2 in each cell of the ovary; fruit a drupe or capsule

99. Burseraceae

b. Ovule 1 in each cell of the ovary; fruit a drupe

105. Anacardiaceae

41a. (39) Plant herbaceous 42

b. Plant woody (tree, shrub or climber) 43

42a. (41) Petals long-clawed, united above the base; leaves not fleshy **117. Stackhousiaceae**

b. Petals entirely free, not long-clawed; leaves fleshy

91. Zygophyllaceae

43a. (41) Flowers, or at least some of them, functionally unisexual (i.e. apparent anthers not producing pollen or ovary containing no ovules) 44

43b. Flowers functionally bisexual 45

44a. (43) Leaves alternate; ovary with 2–5 carpels, not flattened
98. Simaroubaceae

b. Leaves opposite; ovary with 2 (rarely 3) carpels, usually flattened **106. Aceraceae**

45a. (43) Leaves entire or toothed; stamens 4–5, emerging from the disc; seeds with arils **115. Celastraceae**

b. Combination of characters not as above 46

46a. (45) Leaves without stipules, not fleshy; filaments of the stamens united into a tube **100. Meliaceae**

b. Leaves with stipules, fleshy; filaments of the stamens free
91. Zygophyllaceae

47a. (36) Plant herbaceous 48

b. Plant woody (tree, shrub or climber) 50

48a. (47) Leaves always simple; ovary 6–10-celled by the development of 3–5 secondary septa between the original septa during maturing of the flower **92. Linaceae**

b. Leaves lobed or compound; secondary septa absent 49

49a. (48) Leaves with stipules **89. Geraniaceae**

b. Leaves without stipules **88. Oxalidaceae**

50a. (47) Filaments of the stamens united below 51

b. Filaments of the stamens entirely free from each other 55

51a. (50) Plants succulent, spiny; stamens 8 with woolly filaments; plants unisexual **25. Didiereaceae**

b. Combination of characters not as above 52

52a. (51) Stamens 2 **179. Oleaceae**

b. Stamens 3 or more 53

53a. (52) Leaves without stipules **13. Olacaceae**

b. Leaves with stipules though these sometimes quickly deciduous 54

54a. (53) Stipules persistent, born between the bases of the leafstalks; petals with appendages **93. Erythroxylaceae**

b. Stipules quickly deciduous, not borne as above; petals without appendages **127. Sterculiaceae**

55a. (50) Stamens 8–10 56

b. Stamens 3–6 58

56a. (55) Petals long-clawed, often fringed or toothed, stamens 10;

some or all of the sepals with nectar-secreting appendages
101. Malpighiaceae
  b. Petals neither clawed nor fringed or toothed; stamens 8–10; sepals without such appendages   57
57a. (56) Ovule 1 per cell; sepals united at their bases
13. Olacaceae
  b. Ovules many per cell; sepals free   **132. Stachyuraceae**
58a. (55) Staminodes present in flowers which also contain fertile stamens   59
  b. Staminodes absent from flowers with fertile stamens, present only in unisexual (female) flowers   60
59a. (58) Carpels 2, ovary containing a single apical ovule; stipules present, borne between the bases of the leaf-stalks
114. Corynocarpaceae
  b. Carpels 2–4, each cell of the ovary containing 1–2 ovules; stipules absent   **112. Cyrillaceae**
60a. (58) Trees with opposite, pinnate leaves; twigs tipped with large, dark buds; fruit a samara   **179. Oleaceae**
  b. Combination of characters not as above   61
61a. (60) Sepals united at the base   62
  b. Sepals entirely free from one another   63
62a. (61) Carpels 1 or rarely 3 with 1 or 2 of them sterile, the fertile containing 2 apical ovules   **119. Icacinaceae**
  b. Carpels 3–many, all fertile, each containing 1–2 ovules
113. Aquifoliaceae
63a. (61) Ovules 1 per cell; petals 3–4   **97. Cneoraceae**
  b. Ovules many per cell; petals 5   **78. Pittosporaceae**

GROUP F

  1a. Sepals, petals and stamens borne on a rim or cup which is inserted below the ovary   2
  b. Sepals, petals and stamens inserted individually below the ovary   4
  2a. (1) Trees; leaves bi- or tri-pinnate; flowers bilaterally symmetric; stamens 5, of differing lengths   **70. Moringaceae**
  b. Combination of characters not as above   3
  3a. (2) Flower-stalks slightly united to the leaf stalks so that the

flowers appear to be borne on the latter; petals each overlapping 1 other and overlapped by 1 other in bud; carpels 3

133. Turneraceae

b. Flower-stalks not united to the leaf-stalks; petals not so arranged in bud; carpels 2 or 4   **76. Saxifragaceae**

4a. (1) Perianth bilaterally symmetric   5

b. Perianth radially symmetric   9

5a. (4) Ovary of a single carpel with marginal placentation

84. Leguminosae

b. Ovary of 2 or more carpels with parietal placentation   6

6a. (5) Ovary open at the apex; some or all of the petals fringed

69. Resedaceae

b. Ovary closed at the apex; no petals fringed   7

7a. (6) Petals and stamens 5; carpels 3   **131. Violaceae**

b. Petals and stamens 4 or 6; carpels 2   8

8a. (7) Ovary borne on a stalk (gynophore); stamens projecting beyond the petals   **67. Capparaceae**

b. Ovary not borne on a stalk; stamens not projecting beyond the petals   **66. Papaveraceae**

9a. (4) Petals and stamens numerous   **19. Aizoaceae**

b. Petals and stamens each fewer than 7   10

10a. (9) Stamens alternating with much-divided staminodes

76. Saxifragaceae

b. Stamens not alternating with much-divided staminodes   11

11a. (10) Leaves insect-trapping and digesting by means of stalked, glandular hairs   **65. Droseraceae**

b. Leaves not as above   12

12a. (11) Climbers with tendrils; ovary and stamens borne on a common stalk (androgynophore); corona present

134. Passifloraceae

b. Combination of characters not as above   13

13a. (12) Petals 4, the outer pair trifid; sepals 2   **66. Papaveraceae**

b. Petals not as above; sepals 4–5   14

14a. (13) Stamens usually 6, 4 longer and 2 shorter, rarely reduced to 2; carpels 2; fruit usually with a secondary septum

68. Cruciferae

b. Stamens 4–10, all more or less equal; carpels 2–5; fruit without a secondary septum   15

15a. (14) Petals each with a scale-like appendage at the base of the blade; leaves opposite　　　**138. Frankeniaceae**

  b. Petals without appendages; leaves alternate or all basal　　**16**

16a. (15) Stipules present　　**17**

  b. Stipules absent　　**18**

17a. (16) Stamens 10; flowers in dense, cylindric panicles

　　　　　　77. Cunoniaceae

  b. Stamens 5; inflorescence not as above　　**131. Violaceae**

18a. (16) Leaves alternate, scale-like　　**137. Tamaricaceae**

  b. Leaves usually all basal, normally developed　**167. Pyrolaceae**

GROUP G

1a. Aquatic plants or rhubarb-like marsh plants with cordate leaves　　**2**

  b. Terrestrial plants, not as above　　**4**

2a. (1) Stamens 8, 4 or 2; leaves deeply divided or cordate at the base　　**154. Haloragaceae**

  b. Stamen 1; leaves undivided, not cordate at base　　**3**

3a. (2) Leaves in whorls; fruit small, indehiscent, dry, 1-seeded, not lobed　　**156. Hippuridaceae**

  b. Leaves opposite; fruit 4-lobed, with up to 4 seeds

　　　　　　194. Callitrichaceae

4a. (1) Trees or shrubs　　**5**

  b. Herbs, climbers or parasites　　**18**

5a. (4) Plant covered by scales; fruit enclosed in the berry-like, persistent, fleshy calyx　　**129. Eleagnaceae**

  b. Plant not covered in scales; fruit not as above　　**6**

6a. (5) Stamen 1, or 1 complete stamen flanked by 2 half-stamens; leaves opposite　　**49. Chloranthaceae**

  b. Stamens not as above; leaves usually alternate　　**7**

7a. (6) Stamens as many as and on radii alternating with the sepals

　　　　　　120. Rhamnaceae

  b. Stamens not as above　　**8**

8a. (7) Stamens 4, situated at the tops of the spoon-shaped, petal-like perianth-segments which split apart as the flower opens

　　　　　　12. Proteaceae

  b. Combination of characters not as above　　**9**

9a. (8) Ovary 2-celled, partly inferior; stellate hairs often present; fruit a woody capsule **73. Hamamelidaceae**

b. Combination of characters not as above 10

10a. (9) Ovary inferior 11

b. Ovary superior 14

11a. (10) Placentation parietal **76. Saxifragaceae**

b. Placentation axile or basal 12

12a. (11) Styles 3–6; fruit a nut surrounded by a scaly cupule
7. Fagaceae

b. Style 1; fruit not as above 13

13a. (12) Stamens 4–5; placentation basal **14. Santalaceae**

b. Stamens 5–10; placentation axile **161. Cornaceae**

14a. (10) Leaves aromatic, dotted with translucent glands; anthers opening by flaps **36. Lauraceae**

b. Leaves neither aromatic nor gland-dotted; anthers not opening by flaps 15

15a. (14) Stamens numerous, borne on the reflexed inner surface of the perianth **82. Rosaceae**

b. Stamens not as above 16

16a. (15) Stamens 2, or 8–10 borne on different levels in the perianth-tube; leaves simple, entire **128. Thymelaeaceae**

b. Stamens not as above; leaves lobed or compound 17

17a. (16) Inflorescence borne on the shoots of the current year; fruit a schizocarp of 2 (rarely 3) samaras **106. Aceraceae**

b. Inflorescence borne on the older wood; fruit a legume
84. Leguminosae

18a. (4) Plants parasitic 19

b. Plants not parasitic 22

19a. (18) Branch-parasites with green, forked branches or flowers stalkless on branches of host 20

b. Root parasites lacking chlorophyll 21

20a. (19) Flowers borne on green, forked branches
15. Loranthaceae

b. Flowers brown, minute, unstalked **51. Rafflesiaceae**

21a. (19) Flowers minute in fleshy spikes; stamen 1
157. Cynomoriaceae

b. Flowers conspicuous in short, bracteate spikes
51. Rafflesiaceae

22a. (18) Perianth absent; flowers in spikes **47. Saururaceae**
  b. Perianth present; flowers not usually in spikes 23
23a. (22) Leaf-base oblique; ovary inferior, 3-celled
143. Begoniaceae
  b. Leaf-base not oblique; ovary not as above 24
24a. (23) Ovary superior 25
  b. Ovary inferior 30
25a. (24) Carpel 1, containing a single apical ovule; perianth
  tubular **128. Thymelaeaceae**
  b. Combination of characters not as above 26
26a. (25) Carpels 3 (rarely 2), ovule 1, basal; perianth persistent in
  fruit; leaves usually alternate, entire 27
  b. Combination of characters not as above 28
27a. (26) Leaves without stipules; stamens 5 **21. Basellaceae**
  b. Leaves usually with stipules united into a sheath (ochrea);
  stamens usually 6–9 **16. Polygonaceae**
28a. (26) Leaves alternate, usually lobed or compound
82. Rosaceae
  b. Leaves opposite, usually entire 29
29a. (28) Ovule 1, fruit a nut; stipules translucent and papery or
  rarely absent **22. Caryophyllaceae**
  b. Ovules numerous; fruit a capsule; stipules absent
145. Lythraceae
30a. (24) Leaves pinnate; ovary open at apex **142. Datiscaceae**
  b. Leaves not pinnate; ovary closed at apex 31
31a. (30) Ovary 6-celled; perianth 3-lobed or tubular and bilaterally
  symmetric **50. Aristolochiaceae**
  b. Combination of characters not as above 32
32a. (31) Ovules 1–5; seed 1 33
  b. Ovules and seeds numerous 34
33a. (32) Perianth-segments thickening in fruit; leaves alternate
23. Chenopodiaceae
  b. Perianth-segments not as above; leaves opposite or alternate
14. Santalaceae
34a. (32) Styles 2; placentation parietal **76. Saxifragaceae**
  b. Style 1; placentation axile **153. Onagraceae**
GROUP H
 1a. Aquatic herb; leaves divided into thread-like segments
46. Ceratophyllaceae

1b. Terrestrial plants; leaves not as above   2

2a. (1) Trailing, heather-like shrublets; fruit a berry
169. Empetraceae

  b. Combination of characters not as above   3

3a. (2) Flowers in racemes or spikes; fruit a berry or drupe-like; leaves entire, alternate, without stipules; carpels more than 5
17. Phytolaccaceae

  b. Combination of characters not as above   4

4a. (3) Ovary 3-celled; styles 3   5

  b. Ovary 1-, 2- or 4-celled; styles 1–2   7

5a. (4) Leaves with sheathing, membranous stipules; perianth-segments 6; fruit a nut   **16. Polygonaceae**

  b. Combination of characters not as above   6

6a. (5) Fruit a schizocarp; sap often milky; deciduous or evergreen herbs, shrubs or trees or stem-succulents; styles usually divided; seeds usually appendaged   **94. Euphorbiaceae**

  b. Fruit a capsule splitting through the cells; sap not milky; evergreen shrubs; styles undivided; seeds black and shiny, without appendages   **118. Buxaceae**

7a. (4) Resinous trees or shrubs; leaves simple or pinnate; flowers with nectar-secreting disc; stamens 3–10; fruit 1-seeded, drupe-like   **105. Anacardiaceae**

  b. Combination of characters not as above   8

8a. (7) Stamens 2   **179. Oleaceae**

  b. Stamens more than 2   9

9a. (8) Tips of the leaves extended as insect-trapping pitchers
64. Nepenthaceae

  b. Tips of the leaves not as above   10

10a. (9) Plants aromatic, dioecious; stamens 3–18, filaments united; ovary of 1 carpel containing a single, basal ovule
30. Myristicaceae

  b. Combination of characters not as above   11

11a. (10) Placentation parietal; stamens numerous; fruit a berry or capsule   **130. Flacourtiaceae**

  b. Combination of characters not as above   12

12a. (11) Trees, shrubs or woody climbers, if herbaceous then flowers sunk in a fleshy receptacle; ovules apical   13

  b. Combination of characters not as above   16

13a. (12) Ovules 4, of which only 1 develops; flowers in axillary racemes **95. Daphniphyllaceae**
   b. Ovule 1; flowers not in axillary racemes 14
14a. (13) Sap watery; fruit a drupe **8. Ulmaceae**
   b. Sap milky; fruit a syncarp or group of samaras 15
15a. (14) Perianth present; fruit frequently a syncarp of drupes or achenes united with the flat to flask-shaped receptacle
   10. Moraceae
   b. Perianth absent; fruit a group of samaras **9. Eucommiaceae**
16a. (12) Stinging hairs present or plant rough to the touch; stamens touch-sensitive, inflexed in bud; leaves often with cystoliths; seed with a straight embryo **11. Urticaceae**
   b. Stinging or rough hairs absent; stamens neither touch-sensitive nor inflexed in bud; cystoliths absent; seed often with a curved embryo 17
17a. (16) Perianth translucent and papery; stamens usually with the filaments united below **24. Amaranthaceae**
   b. Perianth greenish or absent; filaments free 18
18a. (17) Leaves all opposite; fruit splitting into 2 mericarps
   94. Euphorbiaceae
   b. Leaves alternate, at least above; fruit not as above 19
19a. (18) Male flowers with 7–22 stamens; style at last lateral; leaf-bases sheating **155. Theligonaceae**
   b. Combination of characters not as above 20
20a. (19) Ovary with septa, containing 4 ovules; leaves leathery
   118. Buxaceae
   b. Ovary 1-celled, without septa, containing 1 ovule; leaves not leathery 21
21a. (20) Leaves with stipules; ovule apical **10. Moraceae**
   b. Leaves without stipules (sometimes succulent and continuous with the stem) ovule basal **23. Chenopodiaceae**

GROUP I

   1a. Flowers in racemes or spikes; fruit a berry or drupe-like; leaves entire, alternate, without stipules **17. Phytolaccaceae**
   b. Combination of characters not as above 2
   2a. (1) Trees or trailing, heather-like shrublets, rarely aromatic shrubs 3

2b. Herbs, climbers or non-aromatic shrubs   9

3a. (2) Trailing, heather-like shrublet; fruit a drupe

169. Empetraceae

  b. Trees or aromatic shrubs; fruit a drupe, samara, nut or capsule   4

4a. (3) Stamens numerous; ovary with 5 or more cells   5

  b. Stamens 12 or fewer; ovary with up to 4 cells   7

5a. (4) Leaves in whorls   **38. Trochodendraceae**

  b. Leaves not in whorls   6

6a. (5) Parts of flowers in 4s, flowers in hanging spikes

37. Tetracentraceae

  b. Flowers with parts not in 4s, flowers in cymes   **124. Tiliaceae**

7a. (4) Leaves evergreen with translucent, aromatic glands; anthers opening by flaps   **36. Lauraceae**

  b. Leaves usually deciduous, without translucent, aromatic glands; anthers opening by longitudinal slits   8

8a. (7) Stamens 2; leaf-base not oblique   **179. Oleaceae**

  b. Stamens 4–8; leaf-base oblique   **8. Ulmaceae**

9a. (2) Perianth absent; flowers borne (and often sunk) in a fleshy spike; leaves well-developed, often fleshy   **48. Piperaceae**

  b. Combination of characters not as above   10

10a. (9) Aquatics of running water, resembling algae, mosses or hepatics   **86. Podostemaceae**

  b. Combination of characters not as above   11

11a. (10) Leaves with stipules which are usually united into a sheath (ochrea); fruit usually a 3-sided nut   **16. Polygonaceae**

  b. Leaves without stipules, fruit not a 3-sided nut   12

12a. (11) Sepals falling as soon as the flower opens; herbs with palmately lobed leaves and orange sap   **66. Papaveraceae**

  b. Combination of characters not as above   13

13a. (12) Ovary of 1 carpel; fruit 1-seeded; perianth usually petal-like, bracts sometimes calyx-like   **18. Nyctaginaceae**

  b. Ovary of 2 or more carpels; fruit 1–many-seeded; perianth not petal-like   14

14a. (13) Ovary open at apex; placentation parietal

69. Resedaceae

  b. Ovary closed at apex; placentation basal, free-central or axile   15

15a. (14) Ovule solitary, basal   16
  b. Ovules numerous, axile or free-central   17
16a. (15) Perianth green, membranous or absent; stamens with free
      filaments   **23. Chenopodiaceae**
  b. Perianth translucent and papery; stamens with the filaments
      often united below   **24. Amaranthaceae**
17a. (15) Placentation axile; leaves alternate   **76. Saxifragaceae**
  b. Placentation basal or free-central; leaves usually opposite   18
18a. (17) Sepals free; stamens on the same radii as or more numer-
      ous than perianth-segments   **22. Caryophyllaceae**
  b. Sepals united; stamens as many as and on radii alternating with
      the perianth-segments   **173. Primulaceae**

GROUP J

 1a. Leaves needle-like or scale-like; small, heather-like shrublets
                                                  81. Bruniaceae
  b. Combination of characters not as above   2
 2a. (1) Inflorescence a head subtended by an involucre of bracts;
      ovule always solitary   3
  b. Inflorescence and ovule(s) not as above   4
 3a. (2) Each flower with a cup-like involucel; stamens 4, free;
      ovule apical   **214. Dipsacaceae**
  b. Involucel absent; stamens 5, their anthers united into a tube;
      ovule basal   **219. Compositae**
 4a. (2) Stamens 2; stamens and style united into a touch-sensitive
      column; leaves linear   **218. Stylidiaceae**
  b. Combination of characters not as above   5
 5a. (4) Leaves alternate or all basal   6
  b. Leaves opposite or appearing whorled   14
 6a. (5) Anthers opening by pores; fruit a berry or drupe
                                                  168. Ericaceae
  b. Anthers opening by slits; fruit various   7
 7a. (6) Climbers with tendrils and unisexual flowers; stamens 1–5;
      placentation parietal; fruit berry-like   **144. Cucurbitaceae**
  b. Combination of characters not as above   8
 8a. (7) Stamens 10–many; plants woody   9
  b. Stamens 4–5; plants mainly herbaceous   11

9a. (8) Leaves gland-dotted, smelling of eucalyptus; corolla completely united, unlobed, falling as a whole **147. Myrtaceae**
 b. Combination of characters not as above 10
10a. (9) Hairs stellate or scale-like; stamens in 1 series; anthers linear **177. Styracaceae**
 b. Hairs absent or not as above; stamens in several series; anthers broad **178. Symplocaceae**
11a. (8) Stigma surrounded by a sheath **216. Goodeniaceae**
 b. Stigma not surrounded by a sheath 12
12a. (11) Stamens as many as and on the same radii as the petals **173. Primulaceae**
 b. Stamens not as above 13
13a. (12) Stamens 2 or 4, borne on the corolla; sap not milky **205. Gesneriaceae**
 b. Stamens 5 or more, free from the corolla; sap usually milky **215. Campanulaceae**
14a. (5) Placentation parietal; stamens 2, or 4 and paired **205. Gesneriaceae**
 b. Placentation axile or apical; stamens 1 or more, if 4 then not paired 15
15a. (14) Stamens 1–3; ovary with 1 ovule **213. Valerianaceae**
 b. Stamens 4 or more; ovary usually with 2 or more ovules 16
16a. (15) Leaves divided into 3 leaflets; flowers in a few-flowered head **212. Adoxaceae**
 b. Leaves simple or rarely pinnate; inflorescence various, usually not as above 17
17a. (16) Stipules usually borne between the bases of the leaf-stalks and sometimes looking like leaves; ovary usually 2-celled; flowers usually radially symmetric; fruit capsular, fleshy or schizocarpic **186. Rubiaceae**
 b. Stipules usually absent, when present not as above; ovary usually with 3 cells (occasionally with 2–5 cells), sometimes only 1 cell fertile; flowers often bilaterally symmetric; fruit a berry or drupe **211. Caprifoliaceae**

GROUP K
 1a. Stamens 2 **179. Oleaceae**
 b. Stamens more than 2 2

2a. (1) Carpels several, free; leaves succulent   **74. Crassulaceae**
  b. Carpels united, or, if the bodies of the carpels are more or less free, then the styles united, rarely the ovary of a single carpel; leaves usually not succulent   3
3a. (2) Corolla papery, translucent, 4-lobed; stamens 4, projecting from the corolla; leaves often all basal and with parallel veins
                                                      210. Plantaginaceae
  b. Combination of characters not as above   4
4a. (3) Central flowers of the inflorescence abortive, their bracts forming nectar-secreting pitchers; petals completely united, falling as a whole as the flower opens   **61. Marcgraviaceae**
  b. Combination of characters not as above   5
5a. (4) Stamens more than twice as many as the petals   6
  b. Stamens up to twice as many as the petals   13
6a. (5) Leaves evergreen, divided into 3 leaflets; filaments brightly coloured, at least twice as long as the petals
                                                      60. Caryocaraceae
  b. Leaves deciduous or evergreen, simple, entire or lobed; filaments not as above   7
7a. (6) Leaves with stipules; filaments of stamens united into a tube around the ovary and style   **125. Malvaceae**
  b. Leaves without stipules; filaments free   8
8a. (7) Anthers opening by pores   **56. Actinidiaceae**
  b. Anthers opening by longitudinal slits   9
9a. (8) Leaves with translucent, aromatic glands; calyx cup-like, unlobed   **96. Rutaceae**
  b. Leaves without such glands; calyx not as above   10
10a. (9) Placentation parietal; leaves fleshy   **188. Fouquieriaceae**
  b. Placentation axile; leaves not fleshy   11
11a. (10) Sap milky; ovules 1 per cell   **175. Sapotaceae**
  b. Sap not milky; ovules 2 or more per cell   12
12a. (11) Ovules 2 per cell; flowers usually unisexual
                                                      176. Ebenaceae
  b. Ovules many per cell; flowers bisexual   **59. Theaceae**
13a. (5) Stamens as many as petals and on the same radii as them
                                                      14
  b. Stamens more or fewer than petals, if as many then not on the same radii as them   20

14a. (13) Tropical trees with milky sap and evergreen leaves
175. **Sapotaceae**

　b. Tropical or temperate trees, shrubs, herbs or climbers with watery sap and deciduous leaves　　15

15a. (14) Placentation axile　　16

　b. Placentation basal or free-central　　17

16a. (15) Climbers with tendrils; stamens free　121. **Vitaceae**

　b. Erect shrubs without tendrils; stamens with their filaments united below　　122. **Leeaceae**

17a. (15) Trees or shrubs; fruit a berry or drupe　　18

　b. Herbs (occasionally woody at the extreme base); fruit a capsule　　19

18a. (17) Leaves with translucent glands; anthers opening towards the centre of the flower; staminodes absent　172. **Myrsinaceae**

　b. Leaves without such glands; anthers opening towards the outside of the flower; staminodes 5　171. **Theophrastaceae**

19a. (17) Sepals 2, free　　20. **Portulacaceae**

　b. Sepals 4 or more, united　　173. **Primulaceae**

20a. (13) Flower compressed with 2 planes of symmetry; stamens united in 2 bundles　　66. **Papaveraceae**

　b. Combination of characters not as above　　21

21a. (20) Leaves bipinnate or replaced by phyllodes; carpel 1, fruit a legume　　84. **Leguminosae**

　b. Combination of characters not as above　　22

22a. (21) Anthers opening by pores　　23

　b. Anthers opening by longitudinal slits or pollen in coherent masses (pollinia)　　24

23a. (22) Stamens free from corolla-tube, often twice as many as the petals　　168. **Ericaceae**

　b. Stamens attached to the corolla-tube, as many as the petals
197. **Solanaceae**

24a. (22) Leaves alternate or all basal; carpels never 2 and almost free with a single terminal style　　25

　b. Leaves opposite or whorled, alternate only when carpels 2 and almost free with a single terminal style　　41

25a. (24) Plant woody, leaves usually evergreen; stigma not stalked, borne directly on top of the ovary　113. **Aquifoliaceae**

　b. Combination of characters not as above　　26

26a. (25) Procumbent herbs with milky sap and stamens free from the corolla-tube **215. Campanulaceae**

   b. Combination of characters not as above   27

27a. (26) Ovary 5-celled   28

   b. Ovary 2-, 3- or 4-celled   30

28a. (27) Placentation parietal; soft-wooded tree   **140. Caricaceae**

   b. Placentation axile; herbs   29

29a. (28) Leaves fleshy; anthers 2-celled; fruit often deeply lobed, schizocarpic **196. Nolanaceae**

   b. Leaves leathery; anthers 1-celled; fruit a capsule or berry

    170. Epacridaceae

30a. (27) Ovary 3-celled   31

   b. Ovary 2- or 4-celled   32

31a. (30) Dwarf, evergreen shrublets; 5 staminodes usually present; petals overlapping in bud **165. Diapensiaceae**

   b. Herbs, or climbers with tendrils; staminodes absent; each petal overlapped by and overlapping 1 other in bud

    187. Polemoniaceae

32a. (30) Stamens with the filaments united into a tube; flowers in heads; stigma surrounded by a sheath   **217. Brunoniaceae**

   b. Combination of characters not as above   33

33a. (32) Flowers in spirally coiled cymes or the calyx with appendages between the lobes; style terminal or arising from between the lobes of the ovary   34

   b. Flowers not in spirally coiled cymes, calyx without appendages; style terminal   35

34a. (33) Style terminal; fruit a capsule, usually many-seeded

    190. Hydrophyllaceae

   b. Style arising from the depression between the 4 lobes of the ovary; fruit of up to 4 nutlets, or more rarely a 1–4-seeded drupe

    191. Boraginaceae

35a. (33) Placentation parietal   36

   b. Placentation axile   37

36a. (35) Corolla-lobes edge-to-edge in bud; leaves either of 3 leaflets or simple, cordate or peltate, hairless; aquatic or marsh plants **183. Menyanthaceae**

   b. Corolla-lobes overlapping in bud; leaves never as above; not aquatics or marsh plants **205. Gesneriaceae**

37a. (35) Ovules 1–2 in each cell of the ovary   38
    b. Ovules 3–many in each cell of the ovary   40
38a. (37) Arching shrubs with small purple flowers in clusters on the previous year's wood   **198. Buddlejaceae**
    b. Combination of characters not as above   39
39a. (38) Sepals free; corolla-lobes each overlapping 1 other and overlapped by 1 other, and infolded in bud; twiners, herbs or dwarf shrubs   **189. Convolvulaceae**
    b. Sepals united; corolla-lobes not as above in bud; trees or shrubs   **191. Boraginaceae**
40a. (37) Corolla lobes folded, edge-to-edge or each overlapping 1 other and overlapped by 1 other in bud; septum of the ovary oblique   **197. Solanaceae**
    b. Corolla-lobes variously overlapping but not as above in bud; septum of ovary horizontal   **199. Scrophulariaceae**
41a. (24) Trailing, heather-like shrublet   **168 Ericaceae**
    b. Plant not as above   42
42a. (41) Milky sap usually present; fruit usually of 2 almost free follicles and seeds with silky appendages   43
    b. Milky sap absent; fruit a capsule or fleshy; seeds without silky appendages   44
43a. (42) Pollen granular; corona absent; corolla-lobes each overlapping 1 other and overlapped by 1 other in bud

                                184. Apocynaceae
    b. Pollen usually in coherent masses (pollinia); corona usually present; corolla-lobes edge-to-edge or as above in bud

                           185. Asclepiadaceae
44a. (42) Root-parasite without chlorophyll   **192. Lennoaceae**
    b. Free-living plants with chlorophyll   45
45a. (44) Herbs; flowers in coiled cymes   **190. Hydrophyllaceae**
    b. Herbs or shrubs; flowers not in coiled cymes   46
46a. (45) Placentation parietal; carpels 2   47
    b. Placentation axile; carpels 2, 3 or 5   48
47a. (46) Leaves compound; epicalyx present

                           190. Hydrophyllaceae
    b. Leaves simple; epicalyx absent   **182. Gentianaceae**
48a. (46) Stamens fewer than corolla-lobes   **193. Verbenaceae**
    b. Stamens as many as corolla-lobes   49

49a. (48) Carpels 5; shrubs with leaves with spiny margins
181. Desfontainiaceae

49b. Carpels 2 or 3; herbs or shrubs, leaves not as above   50

50a. (49) Leaves without stipules; carpels 3; corolla-lobes each overlapping 1 other and overlapped by 1 other in bud; plants herbaceous   **187. Polemoniaceae**

  b. Leaves with stipules (often reduced to a ridge between the leaf-bases); carpels usually 2; corolla-lobes variously overlapping or edge-to-edge in bud; plant usually woody   51

51a. (50) Corolla usually 5-lobed; stellate and/or glandular hairs absent   **180. Loganiaceae**

  b. Corolla 4-lobed; stellate and/or glandular hairs present
198. Buddlejaceae

GROUP L

1a. Stamens more numerous than the corolla-lobes, or anthers opening by pores   2

  b. Stamens as many as corolla-lobes or fewer, anthers not opening by pores   6

2a. (1) Anthers opening by pores; leaves undivided; ovary of 2 or more united carpels   3

  b. Anthers opening by longitudinal slits; leaves dissected or compound; ovary of 1 carpel   5

3a. (2) The 2 lateral sepals petal-like; filaments united
103. Polygalaceae

  b. No sepals petal-like; filaments free   4

4a. (3) Shrubs with alternate or apparently whorled leaves; Stamens 5–25   **168. Ericaceae**

  b. Herbs with opposite leaves; stamens 5   **182. Gentianaceae**

5a. (2) Leaves pinnate or of 3 leaflets; perianth not spurred
84. Leguminosae

  b. Leaves laciniate; perianth spurred   **41. Ranunculaceae**

6a. (1) Stamens as many as corolla-lobes; bilateral symmetry weak   7

  b. Stamens fewer than corolla-lobes; bilateral symmetry pronounced   12

7a. (6) Stamens on the same radii as the petals; placentation free-central   **173. Primulaceae**

b. Stamens on radii different from those of the petals; placentation
axile   8

8a. (7) Leaves of 3 leaflets, with translucent, aromatic glands;
stamens 5, the upper 2 fertile, the lower 3 sterile   **96. Rutaceae**

b. Combination of characters not as above   9

9a. (8) Ovary of 3 carpels; ovules many   **187. Polemoniaceae**

b. Ovary of 2 carpels; ovules 4 or many   10

10a. (9) Flowers in coiled cymes; fruit of up to 4 1-seeded nutlets
      191. Boraginaceae

b. Flowers not in coiled cymes; fruit a many-seeded capsule   11

11a. (10) Corolla-lobes each overlapping 1 other and overlapped by
1 other in bud; stamens 5, equal; leaves opposite; climber
      180. Loganiaceae

b. Corolla lobes overlapping in bud but not as above; stamens 4,
or 5 and unequal; leaves usually alternate
      199. Scrophulariaceae

12a. (6) Placentation axile; ovules 4 or many   13

b. Placentation parietal, free-central, basal or apical; ovules many
or 1–2   20

13a. (12) Ovules numerous but not in vertical rows in each cell   14

b. Ovules 4, or more numerous but then in vertical rows in each
cell   16

14a. (13) Seeds winged; mainly trees, shrubs or climbers with opposite, pinnate, digitate or rarely simple leaves **201. Bignoniaceae**

b. Seeds usually wingless; mainly herbs or shrubs with simple
leaves   15

15a. (14) Corolla-lobes variously overlapping in bud; septum of
ovary horizontal; leaves opposite or alternate
      199. Scrophulariaceae

b. Corolla-lobes usually folded, edge-to-edge or overlapped by 1
other and overlapping 1 other in bud; septum of ovary oblique;
leaves alternate   **197. Solanaceae**

16a. (13) Leaves all alternate, usually with blackish, resinous
glands; plants woody   **208. Myoporaceae**

b. At least the lower leaves opposite or whorled, none with glands
as above; plant herbaceous or woody   17

17a. (16) Fruit a capsule; ovules 4–many, usually in vertical rows in
each cell   18

b. Fruit not a capsule; ovules 4, side-by-side   **19**

18a. (17) Leaves all opposite, often prominently marked with cysto-liths; flower-stalks without swollen glands at the base; capsule usually opening elastically, seeds usually on hooked stalks

202. Acanthaceae

  b. Upper leaves alternate, cystoliths absent; flower-stalks with swollen glands at the base; capsule not elastic, seeds not on hooked stalks   **203. Pedaliaceae**

19a. (17) Style arising from the depression between the 4 lobes of the ovary, or if terminal then corolla with a reduced upper lip; fruit usually of 4 1-seeded nutlets; calyx and corolla often 2-lipped   **195. Labiatae**

  b. Style terminal; corolla with well developed upper lip; fruit usually a berry or drupe; calyx often more or less radially symmetric   **193. Verbenaceae**

20a. (12) Ovules 4–many; fruit a capsule, rarely a berry or drupe   **21**

  b. Ovules 1–2; fruit indehiscent, often dispersed in the persistent calyx   **27**

21a. (20) Ovules 4, side-by-side   **193. Verbenaceae**

  b. Ovules many   **22**

22a. (21) Placentation free-central; corolla spurred

207. Lentibulariaceae

  b. Placentation parietal or apical; corolla not spurred, rarely swollen at base   **23**

23a. (22) Leaves scale-like, never green; root parasites   **24**

  b. Leaves green, expanded; free-living plants   **25**

24a. (23) Placentas 4; calyx laterally 2-lipped **206. Orobanchaceae**

  b. Placentas 2; calyx 4-lobed   **199. Scrophulariaceae**

25a. (23) Seeds winged; mainly climbers with opposite, pinnately divided leaves   **201. Bignoniaceae**

  b. Combination of characters not as above   **26**

26a. (25) Capsule with a long beak separating into 2 curved horns; plant sticky-velvety   **204. Martyniaceae**

  b. Capsule without beak or horns; plant velvety or variously hairy or hairless   **205. Gesneriaceae**

27a. (20) Flowers in heads surrounded by an involucre of bracts; ovule 1   **200. Globulariaceae**

  b. Flowers not in heads as above, often in spikes; ovules 1 or 2   **28**

28a. (27) Fruits deflexed; calyx with hooked teeth; ovary 1-celled
    with 1 basal ovule   **209. Phrymaceae**
  b. Fruits mostly erect; calyx without hooks; ovary 2-celled with a
    solitary ovule borne apically in each cell; fruit often 1-seeded
                                                  199. Scrophulariaceae

GROUP M

1a. Trees, shrubs or prickly scramblers with large, pleated usually
    palmately or pinnately divided leaves; flowers more or less
    stalkless in fleshy spikes or panicles with large basal bracts
    (spathes)   **246. Palmae**
  b. Combination of characters not as above   2
2a. (1) Totally submerged aquatic plants of fresh or saline water
                                                                    3
  b. Terrestrial or epiphytic plants, if aquatic then not submerged,
    sometimes entirely floating   6
3a. (2) Plants of fresh or brackish water; flowers bisexual in
    axillary spikes with a perianth of 4 segments edge-to-edge in
    bud and 4 free carpels, *or* marine plants with densely fibrous
    rhizomes (often washed up on beaches as fibre-balls) with leaves
    mostly basal and bisexual flowers in stalked spikes subtended by
    reduced leaves   **206. Potamogetonaceae**
  b. Combination of characters not matching either of the above 4
4a. (3) Flowers on flattened axes or in 2-flowered spikes not
    enclosed by a leaf-sheath; plants of brackish or saline water
                                                  226. Potamogetonaceae
  b. Flowers axillary, not enclosed by the sheaths; plants of fresh or
    brackish water   5
5a. (4) Leaves toothed; ovary solitary with 2–4 slender stigmas
                                                       228. Najadaceae
  b. Leaves entire; ovary of 1–9 free carpels, the stigmas dilated or
    2–4-lobed   **227. Zanichelliaceae**
6a. (2) Small, usually floating aquatic plants not differentiated into
    stem and leaves   **248. Lemnaceae**
  b. Plants various, rarely floating aquatics, plant body differen-
    tiated into stem and leaves   7
7a. (6) Perianth entirely hyaline or papery or reduced to bristles,
    hairs, narrow scales, or absent   8

7b. Perianth well developed, though sometimes small, never entirely hyaline or entirely papery   15

8a. (7) Flowers in small, 2-sided or cylindric spikelets provided with overlapping bracts (spikelets sometimes 1-flowered)   9

b. Flowers arranged in heads, superposed spikes, racemes, panicles or cymes, never in spikelets as above   10

9a. (8) Leaves alternate, in 2 ranks on a stem which is usually hollow and with cylindric internodes; leaf-sheath usually with free margins, at least in the upper part; flowers arranged in 2-sided spikelets (sometimes 1-flowered) each usually subtended by 2 sterile bracts (glumes); each flower usually enclosed by a lower lemma and an upper palea (sometimes absent); perianth of 2–3 concealed scales (lodicules), more rarely of 6 scales or absent; styles generally 2, feathery   **245. Gramineae**

b. Leaves usually arranged on 3 sides of the cylindric or more usually 3-angled stems which usually have solid internodes; young leaf-sheath closed, though sometimes splitting later; flowers arranged in 2-sided or cylindric spikelets, often with a 2-keeled or 2-lobed glume at the base; each flower subtended only by a glume; perianth of several bristles, hairs or scales, or absent; style 1 with 2 or 3 papillose stigmas   **252. Cyperaceae**

10a. (8) Dioecious trees or shrubs with stiffly leathery, sharply toothed leaves, plants often supported by stilt-roots; fruits compound, often woody   **249. Pandanaceae**

b. Combination of characters not as above   11

11a. (10) Inflorescence a simple, fleshy spike (spadix) of inconspicuous flowers, subtended or rarely joined to a large bract (spathe); leaves often net-veined or lobed (plant rarely a small, evergreen, floating aquatic)   **247. Araceae**

b. Combination of characters not as above   12

12a. (11) Flowers unisexual in heads surrounded by an involucre of bracts; perianth in 2 series, often greyish white

244. Eriocaulaceae

b. Combination of characters not as above   13

13a. (12) Flowers bisexual; perianth-segments 6, hyaline or brownish; ovary with 3–many ovules   **239. Juncaceae**

b. Flowers unisexual; perianth-segments a few threads or scales; ovary with 1 ovule   14

14a. (13) Flowers in 2 superposed, elongate, brownish or silvery spikes; ovary borne on a stalk with hair-like branches

251. Typhaceae

b. Flowers in spherical heads; ovary not stalked

250. Sparganiaceae

15a. (8) Carpels free or slightly united at the base   16

b. Carpels united for most of their length though the styles may be free, or carpel solitary   20

16a. (15) Inflorescence a spike, sometimes bifid; perianth-segments 1–4   17

b. Inflorescence not a spike; perianth-segments 6   18

17a. (16) Stamens 6 or more; carpels 3–6; perianth-segments 1–3, petal-like   **224. Aponogetonaceae**

b. Stamens 4; carpels 4; perianth-segments 4, not petal-like

226. Potamogetonaceae

18a. (16) Placentation marginal or parietal   **221. Butomaceae**

b. Placentation basal or marginal   19

19a. (18) Leaf-sheaths with ligules; flowers in racemes; perianth-segments all similar, sepal-like   **223. Scheuchzeriaceae**

b. Leaf-sheaths without ligules; flowers in whorls, racemes or panicles; perianth differentiated into sepals and petals

220. Alismataceae

20a. (15) All perianth-segments similar   21

b. Perianth-segments of the outer and inner whorls conspicuously different, the former usually sepal-like, the latter petal-like   33

21a. (20) Plants with scapes, with small flowers without bracts in racemes or spikes; perianth sepal-like; ovules 1 per cell, basal

225. Juncaginaceae

b. Plants not as above; perianth usually petal-like; ovules usually more than 1 per cell, rarely basal   22

22a. (21) Inflorescence subtended by an entire, spathe-like sheath; plants aquatic   **237. Pontederiaceae**

b. Inflorescence not as above; plants terrestrial   23

23a. (22) Perianth persistent into fruit, covered with branched hairs; stamens 3; sap usually orange   **231. Haemodoraceae**

b. Combination of characters not as above   24

24a. (23) Aquatic herbs with submerged leaves which are 2-toothed at the apex   **242. Mayacaceae**

24b. Combination of characters not as above   25

25a. (24) Outer perianth-whorl with 1 segment much larger than the others; segments of inner whorl petal-like, yellow; flowers with bracts, in heads   **243. Xyridaceae**

  b. Combination of characters not as above   26

26a. (25) Plant woody, or not woody and bearing rosettes of long-lived, fleshy or leathery leaves at or near ground-level   27

  b. Plant herbaceous, leaves usually not long-lived and in rosettes, if so, then deciduous and not fleshy   31

27a. (26) Leaf-stalk bearing 2 tendrils; leaves net-veined

    229. Liliaceae

  b. Leaf-stalk without tendrils; leaves parallel-veined   28

28a. (27) Leaves very small, scale-like or spiny, their function taken over by flattened or thread-like stems (cladodes) on which the inflorescences are often borne   **229. Liliaceae**

  b. Plant with true leaves; cladodes absent   29

29a. (28) Shrubs or woody climbers with scattered stem-leaves; flowers solitary, usually large and hanging; placentation mostly parietal   **229. Liliaceae**

  b. Combination of characters not as above   30

30a. (29) Leaves leathery and more or less thin, if succulent then leathery or not, with a spine-like or cylindric tip; flowers usually green or whitish, bell- or cup-shaped or with a narrow tube and spreading lobes, often more than 1 flower to each bract

    230. Agavaceae

  b. Leaves succulent, usually without a spine-like or cylindric tip; flowers usually red, yellow or orange, tubular, the lobes scarcely spreading, always 1 to each bract   **229. Liliaceae**

31a. (26) Leaves very small, scale-like or spiny, their function taken over by flattened or thread-like stems (cladodes) on which the inflorescences are often borne   **229. Liliaceae**

  b. Plant with true leaves; cladodes absent   32

32a. (31) Leaves evergreen, clearly stalked; flowers more than 1 to each bract, with a narrow tube as long as or longer than the spreading lobes   **230. Agavaceae**

  b. Leaves deciduous, usually without distinct stalks; flowers of various shapes, rarely as above, always 1 to each bract

    229. Liliaceae

33a. (20) Flowers solitary or in umbels; leaves broad, opposite or in a single whorl near the top of the stem   **229. Liliaceae**

  b. Flowers in spikes, heads cymes or panicles; leaves not as above

34

34a. (33) Stamens 6 or 3–5 with 1–3 staminodes; anthers basifixed; leaves usually borne on the stems, often with closed sheaths, never grey with scales; bracts neither overlapping nor conspicuously toothed   **241. Commelinaceae**

  b. Stamens 6, staminodes 0; anthers dorsifixed; leaves mostly in basal rosettes, often rigid and spiny-margined, when on the stems usually grey with scales, bracts usually overlapping and conspicuously toothed   **240. Bromeliaceae**

GROUP N

  1a. Flowers radially symmetric or weakly bilaterally symmetric; stamens 6, 4 or 3, or rarely many   2

  b. Flowers strongly bilaterally symmetric or asymmetric; stamens 5, 2 or 1 (very rarely 6)   14

  2a. (1) Unisexual climbers with heart-shaped or very divided leaves; rootstock tuberous or woody   **236. Dioscoreaceae**

  b. Combination of characters not as above   3

  3a. (2) Perianth persistent into fruit, variously hairy; sap usually orange   **231. Haemodoraceae**

  b. Combination of characters not as above   4

  4a. (3) Rooted or floating aquatics; stamens 2–12; ovules distributed all over the carpel-walls (diffuse-parietal placentation)

222. Hydrocharitaceae

  b. Terrestrial or marsh plants, or epiphytes; stamens 3 or 6, rarely many; placentation axile or parietal (when ovules restricted to a few rows on the carpel-walls)   5

  5a. (4) Stamens 3, staminodes absent; leaves often sharply folded, their bases overlapping; style-branches often divided

238. Iridaceae

  b. Stamens 6 or 3 plus 3 staminodes; leaves usually not as above; style-branches not divided   6

  6a. (5) Placentation parietal; flowers in an umbel with the inner bracts long, thread-like and hanging   **235. Taccaceae**

6b. Placentation usually axile; inflorescence and bracts not as above
7

7a. (6) Perianth consisting of an outer, calyx-like whorl and an inner, corolla-like whorl; bracts usually overlapping and conspicuously coloured **240. Bromeliaceae**

b. Segments of the perianth not in 2 dissimilar whorls; bracts not as above
8

8a. (7) Ovary half-inferior
9

b. Ovary fully inferior
10

9a. (8) Anthers opening by pores **231. Haemodoraceae**

b. Anthers opening by slits **229. Liliaceae**

10a. (8) Leaves long-persistent, evergreen
11

b. Leaves dying down annually
12

11a. (10) Leaves fleshy or leathery, thick, rigid or flexible, spine-tipped, often with spines or teeth on the margins

230. Agavaceae

b. Leaves not spine-tipped and without spines on their margins, somewhat leathery **234. Velloziaceae**

12a. (10) Flowers in a spike; leaves fleshy, often spotted with brown, the margins more or less rolled around each other in bud

230. Agavaceae

b. Flowers in umbels or solitary; leaves not usually fleshy or spotted with brown, but flat, pleated or with the margins folded outwards in bud
13

13a. (12) Leaves mostly basal, densely hairy, pleated or with prominent veins **233. Hypoxidaceae**

b. Leaves various, not usually densely hairy, pleated or with prominent veins, basal or not **232. Amaryllidaceae**

14a. (1) Fertile stamens 6; perianth-segments all similar, tube curved and unevenly swollen; stem below ground, fleshy

230. Agavaceae

b. Stamens 5, 2 or 1, very rarely 6; staminodes, which may be petal-like, often present; perianth-segments usually differing among themselves; fleshy underground stems rare
15

15a. (14) Fertile stamens 2 or 1, united with the style to form a column; pollen usually borne in coherent masses (pollinia); leaf-veins, when visible, all parallel to margins

257. Orchidaceae

15b. Fertile stamens 5 or 1, rarely 6, not united with the style; pollen granular; leaves with a distinct midrib more or less parallel to the margins, the secondary veins parallel, running at an angle from midrib to margins   16

16a. (15) Fertile stamens 5 or rarely 6   **253. Musaceae**

  b. Fertile stamen 1, the remainder transformed into petal-like staminodes   17

17a. (16) Fertile stamen with normal structure, not petal-like

254. Zingiberaceae

  b. Fertile stamen in part petal-like and with only 1 pollen-bearing anther-lobe   18

18a. (17) Leaf-stalk with a swollen band at the junction with the blade; ovary smooth, with 1–3 ovules   **256. Marantaceae**

  b. Leaf-stalk without a swollen band at junction with blade; ovary usually warty, with numerous ovules   **255. Cannaceae**

Arrangement and description of families

Considerations of space have allowed us to give only very short, telegraphic family descriptions. Reference should be made to p. ix for a list of the abbreviations used. In general, the variation given in the descriptions is somewhat wider than that presented in the key, and many characters used in the latter have had to be omitted. Only the families keyed out in the main key are described here; some segregate families are keyed out under the relevant major family.

No attempt has been made to diagnose the orders in which the families are grouped, but for each we have attempted to cite a few differential features, most of which are repeated in the descriptions of the families. The circumscription and content of the orders remain largely matters of opinion.

In using the descriptions, the following features must be assumed for most species of a family, unless otherwise stated: milky sap absent, habit not succulent, parts of the flower free from each other, stamens not antipetalous and anthers opening by longitudinal slits.

The following points concerned with presentation should be noted.

Morphology

The oblique stroke (/) is used instead of 'or'; the letter 'n' is used instead of 'many' (i.e. more than 10 or 12). K – number of calyx segments, C – number of corolla segments, P – number of perianth segments when these are undifferentiated, A – number of stamens, G – number of carpels. These letters are also used in the collective sense, e.g. 'A antipet' means 'stamens antipetalous'. Brackets are used to indicate that the segments of any particular whorl are united to each other, e.g. C(5) means a corolla of 5 united segments.

In the Dicotyledons we have usually not indicated whether the ovary is superior or inferior; however, we have always indicated whether the perianth and stamens are hypogynous, perigynous or epigynous, which gives the same information (see pp. 4–10). Many of the families with petals united into a tube are described as 'K hypog, CA perig'; this means that the stamens are borne on the corolla and the ovary is superior. In the Monoctyledons the ovary position is always indicated.

In plants with inferior ovaries, the number of calyx segments is shown as free if the segments are completely free above their point of attachment on the ovary, and united if they are united above this point. This information is very difficult to obtain.

Information about inflorescence-type is always given, but this, again, is difficult to obtain and condense; the information given here should not be regarded as a complete description of the range of inflorescence-types found in any particular family. We have referred to six main types: racemes, spikes, panicles, cymes, umbels and heads; clusters or fascicles have also been used when the precise nature of the inflorescence is not easily understood. 'Racemose' has been used when spikes, racemes or basically indeterminate panicles occur in the same family; 'cymose' has been used in a similar way to cover various types of determinate inflorescence.

Ovule number refers to the number of ovules in each free carpel (if the ovary is made up of free carpels) or to the ovary as a whole (when it is made up of united carpels), unless qualified by 'per cell'.

Geography

This is indicated in two ways: families with species native to Europe are marked with an asterisk (*) and those with species native to North America are marked with a dagger (†). A short summary is also given of the overall distribution for each family. Most of the terms used for this are self-explanatory; a few, however, need a little explanation. 'Trop' is given for families which occur in both the tropics and subtropics; families occurring only in the subtropics are cited as 'subtrop'. 'Temp' includes the temperate regions of both hemispheres, unless qualified by N or S; in most cases Arctic and Antarctic zones are covered by 'Temp'. We have used 'America' for the whole of the American continent and associated islands; 'Old World' has been used for the eastern hemisphere.

General

For monogeneric families we have given the name of the single genus in parentheses at the end of the description. Some synonyms and short comments about the relationships of some families are also included.

The permitted alternative family names (for the eight families whose traditionally used names do not end in '-aceae') are given, separated from the more familiar name by an oblique stroke, e.g. 164. Umbelliferae/Apiaceae.

Family descriptions

Subclass Dicotyledones

CASUARINALES
Woody, monoecious; branches whip-like, lvs reduced.

1. *Casuarinaceae*
Woody, branches jointed. Lvs whorled, scale-like. Infl catkin. Fls unisex, P0 (?), A1, G(2), naked; ov 2, par. Samaras in woody cones. *Australasia to Malaysia (Casuarina).*

JUGLANDALES
Lvs pinnately compound; infl catkin; monoecious.

2. *Myricaceae**†
Woody. Lvs alt, entire/divided, exstip, aromatic, gland-dotted. Infl catkin. Fls usu unisex. P0, A2–20, usu 4–8, G(2), naked; ov 1, basal. Drupe. *N Hemisphere.*

3. *Juglandaceae**†
Woody. Lvs usu opp, pinnately compound, exstip. At least male fls in catkins. Fls unisex, P4, A3–n, G(2–3), inf; ov 1, basal. Nut with complex lobed and folded cotyledons. *N Temp.*

LEITNERIALES
Dioecious shrubs, catkinate; fr drupe.

4. *Leitneriaceae*†
Woody. Lvs alt, entire, exstip. Infl catkin. Fls variously interpreted; male: P0, A3–12; female: P3–8, G1–loc, sup; ov 1, lateral. Drupe. *USA (Leitneria).*

SALICALES
Dioecious, woody; ov par; fr capsule; seeds woolly.

5. *Salicaceae**†
Woody, dioecious. Lvs usu alt, simple, stip. Infl catkin. Fls usu unisex with disc or nectary gland. P0, A2–n/(2–n), G(2–4), sup; ov n, par. Capsule. Seeds woolly. *Widespread.*

FAGALES

Woody, monoecious; lvs stip; at least male infl usu catkin; G inf/naked; fr nut.

6. Betulaceae*†

Woody. Lvs alt, simple, stip. At least male infl catkin. Fls unisex; male: P(4)/0, A2/4/2–20; female: P0/lobed, G(2), naked/inf; ov 1 per cell, ax/apical. Nut, either small, often winged and in 'cones' or larger, clasped in a bract/cupule. *N Temp.*

Often divided into 2 families:

1a. Nuts small, borne in 'cones'; perianth present in male flowers, absent in female; ovary naked (*Mainly N Temp*)

Betulaceae

b. Nuts larger, subtended by a leaf-like bract or involucre (cupule); perianth present in female flowers, absent in male; ovary inferior (*N Temp*) **Corylaceae**

7. Fagaceae*†

Woody. Lvs usu alt, simple, stip. Male infl often a catkin. Fls unisex; male: P(4–7), A4–n; female: P(4–7), G(3–6), inf; ov 2 per cell, ax. Nut enveloped in cupule. *Temp & Trop.*

URTICALES

Lvs simple, alt; fls small, often unisex, without petals; ovary usu sup; ov 1.

8. Ulmaceae*†

Woody. Lvs alt, simple, stip, usu with oblique base. Infl var. Fls uni/bisex, zyg. PA hypog. P(4–8), A4–8, G(2); ov 1, apical. Samara/drupe. *N Hemisphere.*

9. Eucommiaceae

Trees with milky sap. Lvs alt, simple, exstip. Fls solit, unisex, act. P0, A4–10, G(2) naked; ov 1, apical. Samara. *China (Eucommia).*

10. Moraceae*†

Woody/herbs/climbers, often with milky sap. Lvs alt/opp, simple/divided, stip. Infl var, fls often sunk in expanded receptacle. Fls unisex. PA hypog. Male: P2–6, usu 4/(5), A1–5; female: P2–6 or entire & enveloping ovary, G(2), often 1 carpel aborting; ov 1, apical. Achene/syncarps. *Trop & N Temp.*

Often divided into 2 families:

1a. Usually woody plants with milky sap; perianth in male flowers

of 2–6 free segments; fruit usually a syncarp (*Mostly Trop*)
Moraceae

b.   Herbs without milky sap; perianth in male flowers of 5 united
    segments; fruit an achene (*N Temp*)   **Cannabaceae**

11. *Urticaceae**†

Usu herbs, often with rough or stinging hairs. Lvs alt/opp, simple, usu
stip. Infl var. Fls unisex, act. PA hypog/epig. P0–5/(2–5), A3–5 usu 4,
inflexed in bud and touch-sensitive, G1; ov 1, basal. Achene/drupe.
Widespread.

PROTEALES

Woody; apetalous; P usu 4; G1.

12. *Proteaceae*

Usu woody. Lvs alt, exstip. Infl var. Fls usu bisex, act/zyg. PA perig.
P(4), A4 borne on petal-like, spoon-shaped P segments, G1; ov 1–n,
marginal. Follicle/nut/drupe. *S Hemisphere.*

SANTALALES

Often parasitic, usually green; C absent; G inf.

13. *Olacaceae*

Woody, sometimes half-parasitic on the roots of other trees. Lvs alt,
simple, entire, exstip. Infl cymes/racemes/clusters/solit. Fls usu bisex,
act. KCA hypog/rar epig. K4–6, C3–6/(3–6), some often bilobed,
A6–10, G(3) usu 1-celled; ov 2–3, ax. Drupe. *Tropics.*

14. *Santalaceae**†

Habit var, some half-parasites. Lvs opp/alt, simple, exstip. Fls uni/
bisex, act. PA epig. P4–5/(4–5), A4–5, G(3–5); ov 1–5, basal.
Nut/drupe. *Widespread.*

15. *Loranthaceae**†

Mostly branch-parasites. Lvs usu opp. simple, exstip, leathery. Fls
uni/bisex, act. PA epig. P4–6/(4–6), A4–6, G(3–6); ov not differentiated in flower. Fr a 2–3-seeded berry/drupe. *Mostly Trop.*

POLYGONALES

Stip often prominent and united into a sheath; ov 1, basal.

16. *Polygonaceae**†

Herbs/woody. Lvs usu alt, simple, stip often united into a sheath
(ochrea), rar reduced to a line. Infl often cymose. Fls usu bisex, act. PA

usu hypog. P3–6, A6–9, G(2–4) usu (3); ov 1, basal. Nut. *Mostly N Temp.*

CARYOPHYLLALES
C free/0; embryo usu strongly curved round perisperm.
17. Phytolaccaceae†
Herbs/woody. Lvs alt, entire, exstip. Infl var. Fls usu bisex, act. PA hypog. P4–5/(4–5), A3–n, G1–(n), rar free; ov 1 or 1 per cell, basal/ax. Fr often fleshy. *Mainly American Trop & S Hemisphere.*
18. Nyctaginaceae†
Herbs/woody. Lvs usu opp, entire, exstip. Infl cymose. Fls usu bisex, act. PA hypog. P(5) petal-like, tubular, A(1–30), G1-loc; ov 1, basal. Achene often in persistent P. *Trop.*
19. Aizoaceae*†
Herbs/shrubs, usu leaf-succulents. Lvs usu opp, simple, stip/exstip. Fls bisex, act. KCA hypog/perig/epig. K(5–8), C n, A n, G(3–n); ov usu n, usu par. Fr usu a complex capsule opening when wet, closing when dry. *Mostly South Africa.*
    Including the Molluginaceae.
20. Portulacaceae*†
Herbs/shrubs, often fleshy. Lvs alt/opp, simple, stip. Infl var. Fls bisex, act, KCA hypog/epig. K2/(2), rar 3, C3–12/(3–12) when united only at extreme base, usu 4–6, A3–n, antipet when few, G(2–3); ov 1–n, basal/free-central. Capsule. *Mostly New World.*
21. Basellaceae†
Climbers. Lvs alt, simple, exstip. Infl racemose. Fls bisex, act. PA perig. P5/(5), A5, G(3); ov 1, basal. Fr in persistent, fleshy P. *Mostly trop America.*
22. Caryophyllaceae*†
Herbs, rar shrublets. Lvs usu opp, entire, stip/exstip. Infl cymose/fls solit. Fls usu bisex, act. KCA hypog/perig, rar PA hypog/perig. K4–5/(4–5), C4–5/0, A3–10, G(2–5); ov usu n, free-central or 1 and basal. Capsule/nut rar fleshy. *Mostly N Temp.*
23. Chenopodiaceae*†
Herbs/shrubs. Lvs alt/opp, simple, exstip/stip reduced to scales when stem fleshy and segmented. Infl usu cymose. Fls uni/bisex, act. PA usu hypog. P(3–5) rar 0, green/membranous, A usu 5, G(2–3), rar half-inf; ov 1, basal. Achene/nut. *Widespread.*

24. Amaranthaceae*†

Herbs/woody. Lvs alt/opp, usu entire, exstip. Infl often racemose. Fls usu biscx, act. PA hypog. P3–5/(3–5) usu hyaline and/or papery, A usu (5), G(2–3); ov 1–few, basal. Capsule/achene. *Mostly trop.*

25. Didiereaceae

Woody, succulent, often spiny. Lvs alt, exstip. Cymes/panicles. Fls unisex, act. KCA hypog. K2, persistent, C4 in 2 pairs, A usu 8–10, G(3–4), 1-celled; ov 1, basal. Fr dry, indehiscent. *Madagascar.*

CACTALES

Succulents, often spiny; lvs usu 0; K & C n.

26. Cactaceae†

Mostly spiny stem-succulents. Lvs usu absent. Fls usu solit & bisex, act/slightly zyg. KCA epig. K n, C n/(n), A n, G(3–n); ov n, par. Berry. *Mostly America.*

MAGNOLIALES

Aromatic trees/shrubs/climbers; G1–n, carpels usu free.

27. Magnoliaceae†

Trees/shrubs. Lvs simple, alt, with large, deciduous stip. Fls bisex, act, solit. PA/KCA hypog. P/KC in several series, in 3s or rar 4s, A n, spirally arranged, G n, spirally arranged; ov 2–n, marginal. Fr a group of follicles. Seeds large. *Mostly N Temp & subtrop.*

28. Winteraceae

Woody. Lvs alt, simple, exstip. Fls bisex, act, in cymes/fascicles. KCA hypog. K2–6/(2–6), valvate in bud, C n in 2–several series, A n, G1–n in 1 whorl; ov 1–n, marginal. Follicle/berry. *Tropics (except Africa), S Temp.*

29. Annonaceae†

Woody. Lvs simple, exstip. Infl var. Fls usu bisex, act. KCA hypog. K usu 3, C3–6, A n, each crowned by an enlarged connective, G n, usu stalked in fr, rarely united into a mass; ov 1–n, basal/marginal. Seeds with arils, endosperm convoluted. *Trop.*

30. Myristicaceae

Dioecious trees. Lvs alt, entire, exstip. Infl racemose. Fls unisex, act. PA hypog. P(2–5) usu (3), valvate in bud; male: A(2–20), filaments united; female: G1; ov 1, basal. Fr fleshy, 2-valved. Seed with coloured aril, endosperm convoluted. *Trop.*

31. Canellaceae

Woody. Bark very aromatic. Lvs alt, simple, exstip. Infl cymes/racemes. Fls bisex, act. KCA hypog. K3, C4–5, A more than 10, G(2–6), 1-celled, ov 2–n, par. Berry. *Trop & subtrop Africa & America.*

32. Schisandraceae†

Woody climbers. Lvs alt, simple, exstip. Fls unisex, act, axillary. KCA hypog. K & C 9–15, poorly differentiated, A (n) usu united into a fleshy mass, G n; ov 2–3. Fr berry-like crowded or distant on an elongate axis. *N America, E Asia.*

33. Illiciaceae†

Woody, aromatic. Lvs alt/whorled, simple, exstip. Fls bisex, act, solit. KCA hypog. K & C 7–n, imbricate, A4–n, G5–n in a single whorl; ov 1, almost basal. Fr a group of follicles. *SE Asia, SE North America (Illicium).*

34. Monimiaceae

Trees/shrubs. Lvs opp, simple, exstip, usu evergreen. Fls solit/cymose, usu unisex, act. KCA/PA perig. K4–n, C4–n/P4–n, A usu n, G1–n; ov 1. Achenes in enlarged perigynous cup. *Mainly trop.*

35. Calycanthaceae†

Shrubs. Lvs opp, entire, exstip. Fls solit, bisex, act. PA perig. P n, A5–30, G n; ov 1–2, marginal. Achenes. *SE USA.*

36. Lauraceae*†

Woody. Lvs usu alt, entire, exstip, glandular-punctate. Infl cymose/racemose. Fls small, uni/bisex, act. PA hypog/perig. P usu (6), A12 or variable, anthers opening by valves; G1; ov 1, apical. Drupe-like berry. *Mostly Trop.*

37. Tetracentraceae

Woody. Lvs alt, simple, exstip. Infl catkin-like. Fls bisex, act. PA hypog. P4, A4, G(4); ov n, marginal. Capsule. *China, adjacent Burma (Tetracentron).*

38. Trochodendraceae

Woody. Lvs whorled, simple, exstip. Fls act, bisex, racemose/in clusters. PA hypog. P minute or 0, A n, G(6–10); ov n, marginal. Fr dehiscent. *Japan (Trochodendron).*

39. Eupteleaceae

Woody. Lvs alt, exstip. Fls act, bisex. P0, A n, G6–n, stalked; ov 1–3, marginal. Fr a group of stalked samaras. *Himalaya, Japan, China (Euptelea).*

40. Cercidiphyllaceae

Trees. Lvs deciduous, simple, opp/alt, stip. Fls unisex, plants dioecious; male: almost stalkless, P4, A15–20; female: stalked, P4, hypog, G4–6; ov n, marginal. Follicles. *China, Japan (Cercidiphyllum).*

41. Ranunculaceae*†

Usu herbs/climbers. Lvs usu alt, simple/compound, usu exstip. Infl var. Fls bisex, act/zyg. KCA/PA hypog. P4–n/K3–5, C2–n rar (4), A n, G1–n rar united; ov 1–n, marginal/basal/apical/rar axile. Achenes/follicles. *Mainly temp.*

42. Berberidaceae*†

Herbs/shrubs. Lvs alt. simple/divided, usu exstip. Infl cymose/racemose/fls solit. Fls bisex, act. KCA hypog, K & C sometimes not well differentiated. K4–6, C4–6 rar 9, A4–18 often antipet, anthers often opening by valves, G apparently 1; ov few, basal/marginal. Capsule/berry. *Mainly N Temp.*

43. Lardizabalaceae

Woody, usu climbers. Lvs alt, compound, exstip. Infl raceme-like. Fls usu unisex, act. PA hypog. P3–6/K3–6, C6, A6/(6), G3–15; ov n, marginal. Berries. *Scattered.*

44. Menispermaceae†

Usu woody climbers. Lvs alt, simple, exstip. Fls unisex, act. PA hypog. P3/6/9, A3–n usu 6, G3–6; ov 1 marginal. Drupe/achene. *Mostly trop.*

45. Nymphaeaceae*†

Rhizomatous aquatics. Lvs alt, cordate/peltate. Fls bisex, act, solit. PA/KCA hypog or serially attached to ovary. P n/K3–6, C3–n, A n, G3–n/(3–n); ov few–n on walls of carpels (par). Fr var. *Widespread.*

Often divided into several smaller families:

1a.	Carpels free	2
b.	Carpels more or less united	3
2a.	(1) Carpels sunk individually in a top-shaped receptacle; perianth-segments in several series (*N Temp*) **Nelumbonaceae**	
b.	Carpels not sunk in the receptacle; perianth-segments in 2 series (*Scattered*) **Cabombaceae**	

3a. (1) Ovary inferior; leaves with prickles (*Trop*) **Euryalaceae**
  b. Ovary superior to inferior; leaves without prickles (*Widespread*) **Nymphaeaceae**

46. *Ceratophyllaceae**†

Submerged aquatics. Lvs whorled, much divided, exstip. Fls solit, unisex. PA hypog. P(10–15), A10–20, G1; ov 1, marginal. Nut. *Widespread (Ceratophyllum).*

PIPERALES

Fls inconspicuous, with bracts, often in spikes.

47. *Saururaceae*†

Herbs. Lvs alt, simple, stip. Infl spike/raceme. Fls bisex, act. P0, A6–8, G3–4/(3–4), sup/inf; ov 1–10 per cell, par/ax. Follicle/fleshy capsule. *Scattered.*

48. *Piperaceae*†

Herbs/shrubs. Lvs usu alt, entire, stip. Infl fleshy spike. Fls minute, usu bisex, often sunk in spike. P0, A1–10, G(2–5), sup; ov 1, basal. Small drupe. *Trop.*

49. *Chloranthaceae*

Herbs/woody. Lvs opp, simple, stip. Infl spike/panicle/head. Fls unisex, act; male: P0, A1–3, sometimes made up of a central whole stamen attached to 2 half-stamens; female: P3, epig, G1–loc; ov 1, pendulous. Drupe. *Trop, S Temp.*

ARISTOLOCHIALES

Herbaceous or shrubby, often climbing; lvs alt, exstip; ov ax; capsule. Flowers or leaves bizarre.

50. *Aristolochiaceae**†

Herbs/climbers. Lvs alt, simple, often cordate, exstip. Infl var. Fls bisex, act/zyg. PA epig. P(3), often bizarre and foetid, A6–n, often attached to style, G(4–6); ov n, ax. Capsule. *Mostly trop.*

51. *Rafflesiaceae**†

Root- or branch-parasites lacking chlorophyll. Lvs scale-like/0. Fls unisex, act. PA epig. P4–10/(4–10), A n, G(4–6–8); ov n, par. Berry. *Mostly Old World trop.*

GUTTIFERALES

Usu woody; A mostly n, often united in bundles.

52. Dilleniaceae
Woody, some climbing. Lvs alt, simple, stip/exstip. Infl var. Fls uni/bisex, act. KCA hypog. K5, C5, A n, often united in bundles, G1–n; ov 1–n, marginal. Follicles/berry-like. *Trop, Australasia.*

53. Paeoniaceae*†
Herbs/shrubs. Lvs alt, compound, exstip. Fls usu solit, bisex, act. KCA hypog. K usu 5, differing among each other, C5 or more, A n, G2–8; ov n, marginal. Large follicles. *N Temp (Paeonia).*

54. Crossosomataceae†
Shrubs. Lvs alt, simple. Fls solit, terminal, bisex, act. KCA perig. K5, C5, A15–n, G3–5; ov n, marginal. Follicles; seeds with much divided arils. *Western N America (Crossosoma).*

55. Eucryphiaceae
Woody. Lvs opp, simple/pinnate, stip. Fls solit, bisex, act. KCA hypog. K4, C4, A n, G(5–12) rar (–18); ov n, ax. Capsule. *Chile, Australia–Tasmania (Eucryphia).*

56. Actinidiaceae
Woody. Lvs alt, simple, exstip. Infl cymes/panicles/clusters. Fls uni/bisex, act. KCA hypog. K5, C5/(5), A10–n, anthers often opening by pores, G(3–5); ov n, ax. Berry/capsule. *Tropics.*

57. Ochnaceae
Usu woody. Lvs alt, usu simple, stip. Infl var. Fls bisex, act. KCA hypog. K4–5/(4–5), C4–6, A5–n, anthers opening by pores, G3–15, carpels free, united by common style; ov 1–n per cell, ax. Usu schizocarp, often fleshy. *Mostly trop.*

58. Dipterocarpaceae
Large trees. Lvs alt, entire, evergreen, stip. Infl usu panicle. Fls bisex, act. KCA hypog. K5, 2 usu enlarging in fr, C5, A n, G usu (3); ov usu 2 per cell, ax rar par. Nut. *Trop Asia, Africa.*

   Large tropical trees occasionally seen in cultivation but very unlikely to flower regularly.

59. Theaceae†
Woody. Lvs alt, simple, exstip. Fls solit/in clusters, usu bisex, act. KCA hypog/CA perig. K5, C5 sometimes united to A at the base, A15–n, free or variously united, G(3–5); ov n, ax. *Trop & warm Temp.*

60. Caryocaraceae
Woody. Lvs alt/opp, evergreen, of 3 leaflets, stip/exstip. Infl raceme.

Fls bisex, act. KCA hypog. K5–6/(5–6), C5–6, A n with long, coloured filaments, G(4–20); ov 1 per cell, ax. Drupe/schizocarp. *Trop America.*

61. Marcgraviaceae

Woody, often epiphytic. Lvs alt, simple, exstip. Infl raceme/umbel; some fls sterile, their bracts variously modified into pitcher-like, pouched or spurred nectaries. KCA hypog. K4–5, C4–5/(4–5) falling as a unit when united, A3–n, G(2–n); ov n, par. Capsule/indehiscent. *Trop America.*

62. Guttiferae/Clusiaceae*†

Herbs/woody. Lvs opp, simple, usu exstip, gland-dotted. Infl cymose. Fls uni/bisex, act. KCA hypog. K2–10/(4–5), C3–12, A usu n in bundles, G(3–5); ov 1–n per cell, ax/par. Capsule/berry. *Widespread.*

The family Hypericaceae, consisting of herbs or shrubs with stamens united in 3–5 bundles and styles divided, is often separated.

SARRACENIALES

Lvs forming insectivorous pitchers or with insect-trapping and -digesting hairs and glands, or insect-trapping by folding of the 2 halves of the leaf.

63. Sarraceniaceae†

Herbs. Lvs basal tubular pitchers. Fls solit/racemose, bisex, act. KCA/PA hypog. K4–5, C5–0, A n, G(3–5), style umbrella-shaped; ov n, ax. Capsule. *America.*

64. Nepenthaceae

Herbs/shrubby climbers. Lvs alt, tips prolonged into insectivorous pitchers. Infl raceme/panicle. Fls unisex, act. PA hypog. P2–4, A4–24, G(3–4); ov n, ax. *Mainly Malaysia (Nepenthes).*

65. Droseraceae*†

Herbs. Lvs in rosettes, usu simple, insectivorous. Infl raceme/panicle. Fls bisex, act. KCA hypog. K(4–5), C5, A5–20, G(3–5); ov n, par. Capsule. Pollen in tetrads. *Temp.*

PAPAVERALES

Fls act (often with 2 planes of symmetry)/zyg; ov par; seeds often with arils; milky or coloured latex sometimes present.

66. Papaveraceae*†

Herbs/rar shrubs, usu with milky or coloured sap. Lvs alt/basal, rar opp, simple; divided, exstip. Infl cymose/fls solit. KCA hypog/rar perig. K usu 2–3 rar 4–n, rar united and falling as a whole, falling early; C0–n usu 4, A n rar 4–6, sometimes variously united, G(2–n) rar almost free; ov 1–n, par. Capsule. *Mostly N Temp.*

Often divided into two segregate families (rarely into 3, the third, Hypecoaceae, included here in Fumariaceae).

1a. Sap milky or coloured; stamens usually numerous; nectaries absent (*N Temp & S America*) **Papaveraceae**

b. Sap not milky or coloured; stamens usually 4, often arranged as 2 groups of 2 half-stamens united with and on either side of a whole stamen; nectar present (*N Temp*) **Fumariaceae**

67. *Capparaceae**†
Woody/herbaceous. Lvs alt, simple/compound, stip/exstip. Infl raceme/fls solit. Fls uni/bisex, act/zyg. KCA hypog. K4–8, C4–n rar 0, A4–n, G(2–4) often stalked; ov few–n, par. Capsule/berry/nut. *Trop, warm Temp.*

68. *Cruciferae/Brassicaceae**†
Usu herbs. Lvs usu alt, simple/divided, exstip. Infl usu without bracts, spike/raceme. Fls usu bisex & act. KCA usu hypog. K4, C4, A6 usu 2 shorter and 4 longer, rar 2, 4 or n, G(2); ov 2–n, par. Usu capsule with false septum (replum). *Temp.*

69. *Resedaceae**†
Herbs/shrubs. Lvs alt, simple/divided, stip minute. Infl spike/raceme. Fls uni/bisex, zyg. KCA usu hypog. K4–8, C4–8 rar 2, A3–n, G(2–6), rar 2–6; ov n, par. Capsule/berry/follicles, usu gaping at apex. *N Temp.*

70. *Moringaceae*
Trees. Lvs alt, 2–3-pinnate, exstip. Infl panicle. Fls bisex, zyg. KCA perig. K5, C5, A5 with 3–5 staminodes, anthers 1-celled, G(3); ov n, par. 3-sided capsule. *Mostly trop Old World (Moringa).*

BATALES
Dioecious shrubs; fls in catkins; fr a syncarp of berries.
71. *Bataceae*†

Shrubs. Lvs opp, simple, exstip. Infl catkin. Fls unisex; male: P2, A4–5 with 4–5 staminodes; female: P0, G(4) naked; ov 1 per cell, ascending. Syncarp of berries. *Coasts of America (Batis).*

ROSALES

Woody/herbaceous; lvs often compound; carpels free/united.

72. Platanaceae*†

Trees with exfoliating bark. Lvs alt, lobed, stip. Infl of spherical heads. Fls unisex, act. PA perig/hypog. P3–5/(3–5), A3–7, G5–9; ov 1/rar 2, marginal. Carpels cohering in fr. *N Temp (Platanus)*.

73. Hamamelidaceae†

Woody, often with stellate hairs. Lvs alt. simple/lobed, stip. Infl var. Fls uni/bisex, act/zyg. KCA perig/epig. K(4–5), C0–4–5, A2–8 anthers often opening by valves, G(2); ov 1–2, ax. Woody capsule. *Scattered*.

74. Crassulaceae*†

Usu leaf succulents. Lvs opp/alt, simple, exstip. Infl cymose. Fls bisex, act. KCA hypog. K3–30/(4–5), C3–30/(4–5), A3–n, G3–n rar united below; ov n, marginal. Follicles. *Widespread (except Australia)*.

75. Cephalotaceae

Herbs. Lvs alt in a rosette, modified into stalked, insectivorous pitchers, exstip. Racemes. Fls bisex, act. PA perig. P6, A12, connective swollen, glandular, G6; ov 1 per carpel, basal/marginal. Follicles. *W Australia (Cephalotus)*.

76. Saxifragaceae*†

Herbs/woody. Lvs usu alt, simple/compound, usu exstip. Fls usu bisex & act. KCA perig/epig/rar hypog. K4–5/(4–5), C4–5 rar 0, A usu 3–10, rar as many as 30, G usu (2), often with divergent styles; ov n, ax/par. Capsule/berry. *Widespread*.

A large family, often separated into a number of segregate families:

1a.	Plants woody	2
b.	Plants herbaceous	5
2a.	(1) Leaves of 3 leaflets, stalkless, opposite, evergreen (*Australia*)	**Baueraceae**
b.	Leaves not as above	3
3a.	(2) Stamens 8 or more; leaves usually opposite (*Mainly E Asia & N America*)	**Hydrangeaceae**
b.	Stamens 4–6; leaves alternate	4
4a.	(3) Disc present; leaves usually with gland-tipped teeth (*mainly S America & Australasia*)	**Escalloniaceae**
b.	Disc absent; leaves without gland-tipped teeth (*Temperate N Hemisphere, Andes*)	**Grossulariaceae**

5a.   (1) Stamens not alternating with staminodes (*Mainly North Temperate regions*)   **Saxifragaceae**

  b.   Stamens alternating with staminodes   6

6a.   (5) Ovary spherical; stamens 5; petals 5; staminodes much divided, gland-tipped (*North Temperate regions*)   **Parnassiaceae**

  b.   Ovary 4-sided, cylindric; stamens 4 or 8; petals 4; staminodes simple (*Temperate S America*)   **Francoaceae**

77. Cunoniaceae

Woody. Lvs opp, usu pinnate, stip. Infl. var. Fls usu bisex, act. KCA perig/epig. K3–6/(3–6), C3–5 rar 0, A n, G(2–5); ov n, ax. Capsule/nut. *Mostly S Hemisphere.*

78. Pittosporaceae

Woody. Lvs alt/opp, simple, exstip. Infl var. Fls bisex, usu act. KCA hypog. K5, C5, A5, G(2–5); ov n, ax. Capsule/berry. *Trop & S Temp Old World.*

79. Byblidaceae

Herbs/small shrubs with insect-trapping glandular hairs. Lvs alt, exstip. Racemes. Fls bisex, act. KCA hypog. K5, C5, A5, anthers opening by pores; G(2); ov n, ax. Capsule. *N & W Australia (Byblis).*

80. Roridulaceae

Low shrubs with insect-trapping glandular hairs. Lvs alt, exstip, sometimes divided. Racemes. Fls bisex, act. KCA hypog. K5, C5, A5, anthers opening by pores, G(3); ov 1–n per cell, ax. Capsule. *South Africa (Roridula).*

81. Bruniaceae

Low heather-like shrublets. Lvs alt, needle-like, exstip. Fls solit or aggregated into racemes/panicles/heads, bisex, act. KCA hypog/epig. K5, C5, A5, G(2–3); ov 1–4 per cell, ax. Capsule/indehiscent. *South Africa.*

82. Rosaceae*†

Herbs/woody. Lvs usu alt & stip, simple/compound, often toothed. Infl var. Fls usu bisex & act. KCA perig/epig. K4–5 rar –9 free or united, sometimes with epicalyx; C0–5 rar –9, A n–4 rar fewer, G1–n/(2–5); ov 1–n, ax/marginal/basal. Follicles/achenes/drupes/pomes. *Widespread.*

83. Chrysobalanaceae†

Woody. Lvs alt, stip. Infl var. Fls bisex, act/zyg. KCA perig. K5, C5,

A2–n, G(2–3) often 1 or 2 carpels sterile, 1-celled, often asymmetrically placed in the perigynous zone; ov 2, basal. Drupe. *Mainly tropics.*

84. *Leguminosae/Fabaceae**†

Herbs/woody. Lvs usu alt, stip & pinnately compound, sometimes with 2–3 leaflets or simple. Infl var. Fls usu bisex & zyg. KCA hypog/perig. K usu 5/(5), C5 free or partially or fully united, An–4, free or variously united, G1 rar –15; ov 1–n, marginal. Legume (sometimes indehiscent)/lomentum/nut. *Widespread.*

Often divided into 3 separate families (which are usually considered as subfamilies when the Leguminosae is retained as the family):

1a. Corolla radially symmetric; petals valvate in bud; stamens 4–many; leaves bipinnate, rarely reduced to phyllodes; seeds with U-shaped lateral line (*mainly Tropics and Subtropics; N America*) **Mimosaceae**

b. Corolla zygomorphic (sometimes weakly so); petals overlapping in bud, rarely absent; stamens 10 or fewer; leaves simply pinnate, of 3 leaflets or simple; seeds usually without a lateral line, rarely with a closed line 2

2a. (1) Upper petal interior, or petal 1 or petals absent; seed usually with a straight radicle (*mainly tropical*) **Caesalpiniaceae**

b. Upper petal exterior; seed usually with an incurved radicle (*Widespread*) **Papilionaceae/Fabaceae**

85. *Krameriaceae*†

Shrubs/herbs. Lvs alt, entire, exstip. Infl axillary/racemose. Fls bisex, zyg. KCA hypog/K hypog CA perig. K4–5, C5/(5), the lower pair often modified into glands, A3–4, anthers opening by pores, G1–loc; ov 2, pendulous. Fr 1-seeded, indehiscent, covered with barbed spines. *Mainly trop America (Krameria).*

PODOSTEMALES

Much modified aquatic herbs.

86. *Podostemaceae*†

Aquatics of running water, resembling algae, mosses or hepatics. Lvs alt, simple. Fls bisex, zyg. PA hypog. P2–3/(2–3), A1–4, G(2); ov n, ax. Capsule. *Mainly Trop.*

GERANIALES

Mostly herbs; lvs simple/divided; G superior; dehiscence of fruit often explosive.

87. Limnanthaceae†

Herbs. Lvs alt, divided, exstip. Fls solit, bisex, act. KCA hypog. K3–5, C3–5, A6–10, G3–5, bodies of carpels free, styles united; ov 1 per cell, ascending. Nutlets. *Temp N America.*

88. Oxalidaceae*†

Mostly herbs. Lvs alt/basal, pinnate, palmate or with 3 leaflets, exstip. Infl var. Fls bisex, act. KCA hypog. K5/(5), C5, contorted, A10, G(5), styles free; ov 1–more, ax. Capsule, often explosive. *Widespread.*

89. Geraniaceae*†

Usu herbs. Lvs alt/opp, simple/compound, stip/exstip. Infl var. Fls bisex, act/zyg. KCA usu hypog. K3–5/(3–5), C3–5/(5), A5–15/(5–10), G(3–5), sup, often long-beaked; ov 1–n per cell, ax. Capsule/berry/schizocarp. *Widespread.*

90. Tropaeolaceae

Herbs. Lvs alt/opp, simple/divided, exstip. Fls bisex, zyg, solitary, axillary. KC partly perig, A hypog. K5, C5, A8, G3-lobed, style 1; ov 1 per cell, ax. Schizocarp. *C & S America (Tropaeolum).*

91. Zygophyllaceae*†

Herbs/shrubs. Lvs usu opp, usu compound, stip, often fleshy. Infl cyme/fls solit. Fls usu bisex & act. KCA hypog, disc usu present. K4–5, C4–5, A5–15, G(2–5); ov n, ax. Capsule/drupe-like/schizocarp. *Trop & warm Temp.*

92. Linaceae*†

Herbs/shrubs. Lvs alt/opp, entire, stip/exstip. Infl cyme. Fls bisex, act. KCA hypog. K4–5/(4–5), C3–5, A4–5 rar 20, G(3–5), often 6–10-celled with 3–5 secondary septa; ov 1–2 per cell, ax. Capsule/drupe. *Widespread.*

93. Erythroxylaceae

Woody. Lvs alt, simple, stip. Infl var. Fls bisex, act. KCA hypog. K5, C5, each with an appendage on inner face, A(10), G(3), often only 1 cell developing; ov 1–2 per cell, ax. Fr berry-like. *Trop (mostly America).*

94. Euphorbiaceae*†

Woody/herbs/succulents, milky sap often present. Lvs usu alt & stip, simple/compound. Infl var. Fls unisex, act (sometimes borne in a gland-bearing cup). PA/KCA hypog. P4–5/rar K5/(5), C5/(5), A1–n, G(2–4) usu (3), styles often divided; ov 1–2 per cell, ax. Fr usu schizocarpic, seeds often carunculate. *Widespread.*

95. Daphniphyllaceae

Trees/shrubs. Lvs alt, crowded, entire, exstip, usu evergreen. Infl axillary racemes. Fls unisex, act. PA hypog. Male: P3–8, imbricate, A6–12; female: P0, staminodes few, small/0, G(2) imperfectly 2-celled, styles 1–2, undivided, persistent; ov 2 per cell, pendulous. 1-seeded drupe. *Temp E Asia (Daphniphyllum).*

RUTALES

Mostly woody, often aromatic; A usu twice as many as C.

96. Rutaceae*†

Woody/herbaceous. Lvs alt/opp, simple/compound, exstip, usu aromatic, gland-dotted. Infl var. Fls usu bisex, act. KCA usu hypog, disc usu present. K3–5/(3–5), C3–5/(3–5), rar 0, A3–10, G4–5, rar n, united at least by a common style, often borne on a short stalk; ov 1–n, ax. Fr fleshy/capsular/samara. *Trop, warm Temp.*

97. Cneoraceae*

Shrubs, sometimes with medifixed hairs. Lvs alt, simple, exstip. Infl cyme. Fls bisex, act. KCA hypog, disc 0. K3–4, C3–4, A3–4, G(3–4) on a stalk; ov 1–2 per cell, ax. Schizocarp. *Mediterranean, Canary Is, Cuba.*

98. Simaroubaceae*†

Woody. Lvs alt, simple/compound, exstip. Infl var. Fls usu unisex, act. KCA hypog, disc present. K(3–8), C0–8, A6–14 rar more, G(2–5); ov 2 per cell, ax. Fr var. *Mainly Trop.*

99. Burseraceae†

Woody with aromatic resins. Lvs alt, usu compound, exstip. Infl panicles/fls solit. Fls uni/bisex, act. KCA hypog with disc. K(3–5), C3–5 rar 0, A6–10, G(2–5); ov 2 per cell, ax. Drupe/capsule. *Mainly Trop.*

100. Meliaceae

Trees, wood often scented. Lvs usu alt, mostly pinnate, exstip. Infl cymose panicles. Fls usu bisex, act. KCA hypog with disc. K(4–5), C4–5 rar –8, A(8–10) rar free or fewer/more numerous, G(2–5); ov usu 2 or more per cell, ax. Berry/capsule/drupe. *Mainly Trop.*

101. Malpighiaceae†
Woody, often with medifixed hairs. Lvs usu opp, simple, stip. Infl var. Fls usu bisex, act. KCA hypog. K5, often with 2 glands on the back, C5, petals often fringed, A(10), G(3); ov 1 per cell, ax. Fr var, often winged mericarps. *Mainly Trop America.*

102. Tremandraceae
Shrublets. Lvs usu opp, entire, exstip. Fls solit, bisex, act. KCA hypog. K4–5, C4–5, A8–10, anthers opening by pores, G(2) ov 1–2 rar 3 per cell, ax. Capsule. *Australia.*

103. Polygalaceae*†
Herbs/shrubs. Lvs usu alt, entire, exstip. Infl racemose. Fls bisex, zyg. KCA hypog/K hypog CA perig. K usu 5, lateral pair petal-like, C usu 3, often joined to staminal tube, A(8–10), anthers opening by pores, G usu (2); ov usu 1 per cell, ax. Usu capsule; seeds with arils. *Widespread.*

SAPINDALES
Woody; A rar antipet; disc often present between perianth and ovary.

104. Coriariaceae*
Usu shrubs, branches angular. Lvs opp, entire, exstip. Infl raceme. Fls usu bisex, act. KCA hypog. K5, C5, keeled inside, A10, G5–10; ov 1 per carpel, apical. Achenes surrounded by fleshy C. *Scattered (Coriaria).*

105. Anacardiaceae*†
Woody, resinous. Lvs usu alt, simple/compound, exstip. Infl panicle. Fls usu bisex & act. KCA hypog, disc often present. K(3–5), C3–5 rar 0, A3–10, G usu (3) rar (1–5); ov 1 per cell, apical/basal. Drupe, 1-seeded. *Trop & warm Temp.*

106. Aceraceae*†
Woody. Lvs usu opp, simple/compound, exstip. Fls clustered/racemes, uni/bisex, act. KC perig, A hypog/perig, disc present. K4–5, C4–5 rar 0, A4–5, G(2–3); ov 2 per cell, ax. Winged mericarps. *N Temp.*

107. Sapindaceae†
Usu woody. Lvs usu alt & compound, exstip. Infl var. Fls uni/bisex, act/zyg. KCA usu hypog, disc outside A. K5, C4–5 rar 0, A4–n, often 8, G(3) rar 1(–4); ov 2 per cell, ax. Fr var, seeds often with arils. *Mainly Trop.*

108. Hippocastanaceae*†

Woody. Lvs opp, palmate, exstip. Infl racemose. Fls usu bisex, zyg. KC perig, A usu hypog, disc present. K(4–5), C4–5, A5–9, G(3); ov 2 per cell, ax. Capsule, seeds large. *Trop & Temp.*

109. Sabiaceae

Woody. Lvs alt, simple/compound, exstip. Infl panicles. Fls usu bisex, zyg. KCA hypog, disc small. K3–5/(3–5), C4–5, A3–5, antipet, G(2); ov 2 per cell, ax. Berry. *Mostly Trop.*

110. Melianthaceae

Usu woody. Lvs alt, pinnate, stip between leaf-stalk and stem. Infl racemes. Fls bisex, zyg. KC perig, A hypog, with disc. K5/rar 4, C4–5, A4–5/rar 10, free/united, G(4–5); ov 1–n per cell, ax. Capsule. *Africa.*

111. Balsaminaceae*†

Herbs. Lvs alt/opp, simple, exstip. Infl var. Fls bisex, zyg. KCA hypog. K3/rar 5, often coloured, the lowest spurred; C5, upper exterior, laterals united to each other, A(5), G(5); ov n, ax. Explosive capsule. *Mostly Old World.*

Placed with the Geraniales (p. 87) in most other taxonomic systems.

CELASTRALES

Woody; petals usu present, free; hypogynous disc often well developed.

112. Cyrillaceae†

Woody. Lvs alt, simple, exstip. Infl racemes. Fls bisex, act. KCA hypog, disc 0. K(5) C5/(5), A5/10, G(2–4); ov 1–2 per cell, ax. Fr dry, indehiscent. *Mainly Trop America.*

113. Aquifoliaceae*†

Woody. Lvs alt, simple, exstip. Fls in clusters/cymes, uni/bisex, act. KCA hypog, disc 0. K(3–6), C4–5/(4–5), A4–5, G(3–n); ov 1–2 per cell, ax. Drupes. *Widespread.*

114. Corynocarpaceae

Woody. Lvs alt, evergreen, stip. Racemes/panicles. Fls bisex, act. KCA hypog. K5, C5, A5, antipet, G(2); ov 1, hanging. Drupe. *Australasia, Polynesia.*

115. Celastraceae*†

Woody. Lvs alt/opp, simple, stip/exstip. Infl cymes. Fls usu bisex, act,

KCA hypog/perig with disc. K(5), C4–5 rar 10, A4–5 rar 10, G(2–5); ov usu 2 per cell, ax. Fr var, seeds with arils. *Widespread.*

116. *Staphyleaceae**†

Woody. Lvs usu opp, compound, stip. Infl raceme/panicle. Fls usu bisex, act. KCA perig with disc. K(5), C5, A5, G(2–3); ov n, ax. Inflated capsule. *Trop & Temp.*

117. *Stackhousiaceae*

Herbs/small shrubs. Lvs alt, simple, exstip. Racemes. Fls bisex, act. KCA perig. K5, C5, free at the base sometimes united above, A5, G(2–5); ov 1 per cell, ax/basal. Schizocarp. *Australasia.*

118. *Buxaceae**†

Evergreen, usu woody. Lvs usu opp, simple, exstip. Infl spike/raceme/ clusters. Fls unisex, act. PA hypog. P(4–12) rar 0, A4–n, G(2–4) usu (3), styles undivided; ov usu 2 per cell, ax. Loculicidal capsule/berry-like, seeds shiny black. *Mostly Trop.*

119. *Icacinaceae*

Trees/shrubs/climbers. Lvs alt, exstip, simple. Infl cymose. Fls uni/ bisex, act. KCA hypog. K4–5, C4–5, A4–5, G(2–5); ov 2 per cell, ax, pendulous. Drupe/samaras. *Tropics.*

RHAMNALES

Woody; A antipet; KCA often perig, disc usu present; ov 1–2 per cell.

120. *Rhamnaceae**†

Woody. Lvs usu alt & stip, simple. Infl corymb/cyme/clusters. Fls uni/bisex, act. KCA perig/epig, disc usu present. K4–5, C4–5, A4–5, antipet, G(2–4); ov 1, rar 2 per cell, ax. Capsule/drupe-like. *Trop, N. Temp.*

121. *Vitaceae**†

Usu climbers with tendrils. Lvs alt, simple/compound, stip/exstip. Infl often cyme, leaf-opposed. Fls uni/bisex, act. KCA perig, with disc. K4–5/(4–5), C4–5/(4–5), often falling as a unit, A4–5, antipet, G(2–6) usu (3); ov 1–2 per cell, ax. Berry. *Trop & warm Temp.*

122. *Leeaceae*

Shrubs. Lvs alt, simple/compound, exstip. Panicles. Fls bisex, act. KCA perig, with disc. K4–5/(4–5), C4–5, A(4–5), antipet, G(3–8); ov 1 per cell, ax. Berry. *Old World Tropics (Leea).*

MALVALES

Stellate hairs common; K edge-to-edge in bud; filaments often united into a tube round ovary.

123. Elaeocarpaceae

Woody. Lvs alt/opp, simple, stip. Infl var. Fls usu bisex, act. KCA hypog. K4–5/(4–5), C4–5, often fringed or toothed, A usu n, anthers opening by pores, G(2–n); ov n, ax. Capsule/drupe-like. *Trop.*

124. Tiliaceae*†

Usu woody, often with stellate hairs. Lvs usu alt, simple, stip. Infl cyme. Fls usu bisex, act. KCA hypog. K3–5, usu 5/(5), C0–5, usu 5, A10–n/(10–n), G(4–5); ov 1–n per cell, ax. Fr var. *Widespread.*

125. Malvaceae*†

Herbs/woody, often with stellate hairs. Lvs alt, simple/divided, stip. Infl var. Fls usu bisex, act. K hypog, CA perig. K5/(5), often with epicalyx, C5, contorted, free but united at base to the staminal tube, A (n), anthers 1-celled, G(2–n); ov 1–n per cell, ax. Capsule/schizocarp. *Widespread.*

126. Bombacaceae

Trees often with swollen trunks. Lvs simple/palmate, often scaly, stip deciduous. Fls large, bisex, act. KCA hypog/CA perig. K5/(5), C5, crumpled in bud, A5–n/(5–n), anthers 1-celled, G(2–5); ov 2–n ax. Capsule/indehiscent, seeds often embedded in wool. *Trop.*

127. Sterculiaceae†

Woody, often with stellate hairs. Lvs alt, simple/divided, stip. Infl var. Fls usu bisex & act. KCA hypog. K(3–5), C0–5, A(5–10) rar free, G(4–5); ov 2–n per cell, ax. Fr var. *Mostly Trop.*

THYMELAEALES

Woody; lvs simple, exstip; C usu 0; G1.

128. Thymelaeaceae*†

Usu woody. Lvs alt/opp, simple, exstip. Infl var. Fls bisex, usu act. PA perig. P(4–5)/rar K(4–5), C4, A2–8, G1; ov 1–2, more or less apical. Drupe/nut/capsule. *Widespread.*

129. Elaeagnaceae*†

Woody. Lvs usu alt/opp, simple, exstip, often with scales. Infl var. Fls uni/bisex, act. PA perig. P(2–6) usu (4), A4–12, often alternating with P lobes, G1; ov 1, basal. Achene in fleshy persistent P. *Widespread.*

VIOLALES

Woody/herbaceous; C usu free; G sup/inf; ov usu par.

130. Flacourtiaceae

Woody. Lvs alt, simple, stip. Infl var. Fls usu bisex, act. KCA/PA
hypog/perig/epig. K2–15, C0/2–15, A n/rar 5/rar in antipet bundles,
G(2–n); ov n, par. Capsule/berry. *Mostly Trop.*

131. Violaceae*†

Herbs/shrubs. Lvs usu alt, simple/divided, stip. Infl. var. Fls bisex,
often zyg. KCA hypog. K5, C5, A5, G(3)/rar (5); ov 1–n per cell, par.
Capsule/berry. *Widespread.*

132. Stachyuraceae

Woody. Lvs alt, simple, stip. Infl racemose. Fls usu bisex, act. KCA
hypog. K4, C4, A8, G(4); ov n, ax. Berry. *E Asia (Stachyurus).*

133. Turneraceae

Shrubs/herbs. Lvs alt, simple, exstip. Fls solit/racemose, bisex, act.
KCA perig. K(5), C5, contorted, A5, G(3), stigmas brush-like; ov n,
par. Capsule. *Mainly Trop America.*

134. Passifloraceae†

Often climbers with tendrils. Lvs alt, simple/compound, stip. Fls
axillary, bisex, act. K & C united below, A & G often borne on a stalk
(androgynophore). K4–5/(4–5), C4–5/rar 0, often with corona,
A5/(5), G(3–5); ov n, par. Berry/capsule. *Mainly Trop America.*

135. Cistaceae*†

Herbs/shrubs. Lvs usu opp, simple, stip/exstip. Infl usu cymose/fls
solit/raceme. Fls bisex, act. KCA hypog. K3–5, often differing in size
and shape, C3–5/rar 0, A n, G(3–10) usu (5); ov 2–n per cell, par.
Capsule. *Mainly warm N Temp.*

136. Bixaceae†

Trees/shrubs. Lvs alt, simple, palmately veined/lobed, stip. Infl
panicle/raceme. KCA hypog. K5, imbricate, C5, imbricate, A n,
G(2–5); ov n, par. Capsule. *Trop.*

137. Tamaricaceae*†

Woody. Lvs usu alt, usu scale-like, exstip. Infl usu raceme. KCA
hypog. K4–5, C4–5, A4–10 often arising from a disc, G(3–4); ov n,
par/basal. Capsule, seeds bearded. *Mostly Mediterranean region & C
Asia.*

138. Frankeniaceae*†

Shrubs/herbs. Lvs opp, entire, exstip. Infl cymes/fls solit. Fls bisex, act. KCA hypog. K(4–7), C4–7, each with an outgrowth (ligule) on inner face, A4–7, usu 6, free or united, G(3); ov n, par. Capsule. *Widespread.*

139. Elatinaceae*†

Small usu aquatic herbs. Lvs opp, simple, stip. Fls solit/cymose, bisex, act/zyg. KCA hypog. K3–5/(3–5), C3–5, A6–10, G(3–5); ov n, ax. Capsule. *Widespread.*

140. Caricaceae

Soft-wooded trees/shrubs. Lvs alt, long-stalked, divided, exstip. Infl var. Fls of varying form, often unisex. KCA hypog. K(5), C5/(5), A5–10, G(5); ov n, par. Large berry. *Trop America & Africa.*

141. Loasaceae†

Herbs/shrubs, often with rough/stinging hairs. Lvs opp/alt, simple/ divided, exstip. Fls axillary, bisex, act. KCA usu epig. K4–5, C4–5, A(n), often united in antipet bundles, G(3–7); ov n, ax/par. Capsule. *Mainly America.*

142. Datiscaceae†

Herbs/woody. Lvs alt, simple/compound, exstip. Infl var. Fls unisex, act. PA/KCA epig. P3–8/rar K3–8/(3–8), C0–8, A4–n, G(3); ov n, par. Capsule. *Scattered.*

143. Begoniaceae

Herbs/shrubs. Lvs alt, simple, stip, often fleshy, base often oblique. Fls usu unisex, act/zyg. KCA/PA epig. P2–12/K2, C2, A n/(n), G(2–5) usu (3); ov n, ax. Capsule/berry. *Mostly Trop.*

CUCURBITALES

Herbs, often with tendrils; fr often large.

144. Cucurbitaceae*†

Mostly herbs with tendrils. Lvs alt, often lobed, exstip. Infl axillary cymes/fls solit. Fls usu unisex, act. KCA epig. K5/(5), C5/(5), A1–5 usu 3, 1 anther 1-celled, G(3–5); ov n, par/rar ax. Fr berry-like. *Widespread but mainly Trop.*

MYRTALES

A n–few; G usu inf; ov ax/apical.

145. Lythraceae*†

Usu herbs. Lvs opp, simple, stip/exstip. Fls bisex, usu act. KCA perig. K4–6–8 (epicalyx frequent), C4–8/rar 0, A6–16, rar fewer or more, often with unequal filaments, G(2–6); ov n, ax. Capsule. *Widespread.*

146. *Trapaceae**

Aquatic herb. Lvs opp, simple, exstip, with inflated stalks. Fls bisex, act. KCA more or less epig. K4, C4, A4, G(2); ov 1 per cell. Horned drupe. *Old World (Trapa).*

147. *Myrtaceae**†

Woody. Lvs usu opp, simple, exstip, with translucent aromatic glands. Infl var. Fls bisex, act. KCA usu epig. K4–5/(4–5), C4–5/(4–5), A n/(n), G(3–n); ov 2–n, ax/par. Capsule/berry. *Mostly Trop America & Australia.*

148. *Punicaceae*

Woody. Lvs opp, simple, exstip. Infl cymose/fls solit. Fls bisex, act. KCA epig. K5–8, C5–7, A n, G usu (8–12) rar (3); ov n, ax. Fr berry-like. *Warm N Temp Old World (Punica).*

149. *Lecythidaceae*

Woody. Lvs alt, simple, usu exstip. Fls in large spikes, bisex, act/zyg. KCA epig. K2–6, C4–8/rar 0, A n, variously united, G(2–6); ov 1–n per cell, ax. Fr leathery or woody, seeds large and woody. *Tropics.*

150. *Melastomataceae*†

Woody/herbaceous. Lvs opp, exstip, simple, often with 3 parallel main veins. Infl cymose. Fls bisex, more or less act. KCA perig/epig. K usu 4–5/(4–5), C4–5, A4–10, filaments usu with a conspicuous joint, anthers opening by pores, G usu (4–5); ov n, ax. Capsule/berry. *Mainly Trop.*

151. *Rhizophoraceae*

Trees or shrubs (mangroves). Lvs alt/opp, stip, stipules soon falling. Infl umbel-like/fls solit. Fls usu bisex, act. KCA epig. K3–8 rar more, C3–8, A6–16, G(2–6); ov 1/rar more per cell. Fruit berry-like, seeds often partly developing while still on parent plant. *Trop.*

152. *Combretaceae*

Woody. Lvs opp, entire, exstip. Infl racemose. Fls usu bisex, act. KCA epig. K(4–5)/rar (8), valvate in bud, C4–5/rar 8/0, A4–10/rar n, G1-celled; ov 2–6, apical. Fr leathery, 1-seeded, often winged. *Trop.*

153. *Onagraceae**†

Usu herbs. Lvs alt/opp, simple, stip/exstip. Fls solit/racemose, bisex, usu act. KCA usu epig. K(2–6) usu (4), C4/rar 2, A4–8/rar 1, G(1–5) usu (4); ov n, ax. Capsule/berry/nut. *Mostly Temp.*

154. Haloragaceae*†

Herbs often of damp places. Lvs alt/opp, simple/ divided, stip/exstip. Infl racemose. Fls usu unisex, act. KCA/PA epig. K2–4, C0–4, A2–8, G(2)/(4), sometimes 1-celled; ov 1/1 per cell, ax/apical. Nut/drupe. *Widespread.*

Sometimes separated into 2 families:

1a. Ovary 2–4-celled; stipules absent; leaves deeply dissected (*Widespread*) **Haloragaceae**

 b. Ovary 1-celled; stipules present; leaves entire or lobed (*S Hemisphere*) **Gunneraceae**

155. Theligonaceae*

Fleshy herb. Lower lvs opp, upper alt, all simple and with sheathing bases. Infl cymose. Fls unisex; male: act, PA hypog, P2, A7–22; female: more or less zyg, P tubular, G1, style at last lateral; ov 1, basal. Nut. *Scattered.*

156. Hippuridaceae*†

Aquatic. Lvs whorled, entire, exstip. Fls axillary, bisex. P0, A1, epig, G1-celled; ov 1, apical. Cypsela. *Temp (Hippuris).*

157. Cynomoriaceae*

Root parasites without chlorophyll. Lvs scale-like. Infl spike-like or head-like. Fls usu unisex, act. PA epig. P1–5, A1, G1; ov 1, more or less apical. Small nut. *Mediterranean area, C Asia (Cynomorium).*

UMBELLALES

Flowers with free petals, usu in umbels; G inf; ov ax.

158. Alangiaceae

Trees/shrubs, sometimes spiny. Lvs alt, simple/lobed, exstip. Infl cymes. Fls bisex, act. KCA epig. K(4–10), C4–10, recurving, A4–n, G(2–3), 1-celled; ov 1, pendulous. Drupe. *Old World Trop & Subtrop (Alangium).*

159. Nyssaceae†

Trees. Lvs alt, simple, exstip. Fls in axillary clusters, unisex, act. KCA epig. K(5), C5, A5–12, G1–2-celled; ov 1/1 per cell, ax. Drupe. *Temp N America, China.*

160. Davidiaceae

Trees. Lvs alt, simple, exstip. Infl a head subtended by 2 showy, white bracts, consisting of 1 bisex fl surrounded by many male fls. P0, A1–7, epig in bisex fl, G6–10-celled; ov 1 per cell, ax. Drupe. *China (Davidia)*.

161. Cornaceae* |
Usu woody. Lvs alt/opp, simple, usu exstip. Infl cymes/panicles/heads/rar racemes. Fls uni/bisex, act. KCA epig. K4–5/(4–5), C4–5/rar 0, A4–5, G(2–4); ov 1 per cell, ax/rar par. Drupe/berry. *Mostly Temp*.

162. Garryaceae†
Evergreen shrubs. Lvs opp, entire, exstip. Infl catkin. Fls unisex; male: P4, A4; female: P0–4, G(2), naked/inf; ov 2, apical. Berry. *N America (Garrya)*.

163. Araliaceae*†
Usu woody. Lvs alt, usu lobed/compound, stip, stellate hairs frequent. Infl usu umbel. Fls usu unisex, act. KCA epig. K5/(5), C5–10, valvate in bud, A5–10, G(2–15); ov 1 per cell, ax. Berry/drupe. *Mostly Trop*.

164. Umbelliferae/Apiaceae*†
Usu herbs. Lvs alt, often pinnately compound, stalks sheathing. Infl umbel/rar head. Fls usu bisex, act/zyg. KCA epig. K(5) often very reduced, C5, imbricate and inflexed, A5, G(2); ov 1 per cell, ax. Schizocarp. *Widespread*.

DIAPENSIALES
Shrublets; 5 staminodes present; A opening by slits.
165. Diapensiaceae*†
Usu evergreen shrublets. Lvs alt, simple, exstip. Infl raceme/head/fls solit. Fls bisex, act. K hypog, CA perig. K5/(5), C(5), A5 with 5–0 staminodes, G(3); ov usu n, ax. Capsule. *N Temp*.

ERICALES
Usu woody; lvs simple, exstip, often evergreen; anthers often opening by pores.
166. Clethraceae†
Shrubs/trees. Lvs alt, simple, exstip. Infl raceme/panicle. Fls bisex, act, disc 0. KCA hypog. K(5), C5, A10–12, anthers opening by pores, pollen not in tetrads, G(3); ov n, ax. Capsule. *Mostly Trop & Subtrop (Clethra)*.

167. Pyrolaceae*†

Herbs/shrubs, sometimes saprophytic. Lvs alt, often in rosettes, evergreen, exstip. Fls solit/racemose, bisex, act. KCA hypog. K4–5, C4–5, A8–10, anthers opening by pores, G(4–5); ov n per cell, ax/par. Capsule/berry. *N Temp.*

Often included in Ericaceae.

168. Ericaceae*†

Woody/herbs, rar saprophytic and lacking chlorophyll. Lvs alt/opp/basal, simple, exstip, often evergreen, sometimes needle-like. Infl racemes/clusters/fls solit. Fls bisex, usu act. KCA hypog/epig, rar K hypog, CA perig, disc present. K(4–5)/rar 4–5, sometimes very small, C(3–5)/3–5/rar –10, A5–10 rar –25, anthers usu opening by pores, pollen usu in tetrads, G(2–12); ov n, ax/rar par. Capsule/berry/drupe. *Widespread.*

Formerly, 3 segregate families were recognised within the broad Ericaceae: Ericaceae in the strict sense (ovary superior), Vacciniaceae (ovary inferior) and Monotropaceae (plants saprophytic, without chlorophyll).

169. Empetraceae*†

Shrublets. Lvs alt, entire, exstip, often needle-like. Infl var. Fls usu bisex, act. PA hypog, disc 0. P2–6/rar 0, A2–4, G(2–9); ov 1 per cell, ax. Drupe with 2–9 stones. *Temp.*

170. Epacridaceae

Shrubs. Lvs alt, simple, exstip. Infl racemose. Fls bisex, act. K hypog, CA perig. K(4–5), C(4–5), A4–5, anthers 1-celled, G1–10-celled; ov 1–n per cell, ax. Capsule/drupe. *Mostly Australasia.*

PRIMULALES

Fls usu act; petals usu united; A antipet; ov free-central/basal.

171. Theophrastaceae

Woody. Lvs alt in false terminal whorls, simple, exstip. Infl racemose/clusters/fls solit. Fls uni/bisex, act. K hypog, CA perig. K5/(5), C(5), A5, antipet, anthers opening by pores towards the outside of the flower, with often 5 staminodes; G1-celled; ov n, free-central. Berry/drupe. *New World Trop.*

172. Myrsinaceae

Woody. Lvs alt, simple, exstip, with translucent glands. Infl cymose/clusters. Fls uni/bisex, act. K hypog, CA perig. K4–5/(4–6), C usu

(4–6), A4–6, antipet, opening towards the inside of the flower by slits, G(4–6); ov n usu free-central. Berry/drupe. *Mainly Trop.*

173. *Primulaceae**†

Herbs/rar shrublets/aquatic. Lvs alt/opp/basal, usu simple, exstip. Infl var. Fls bisex, usu act. K hypog, CA perig. K(4–9) usu (5), C(4–9) usu (5)/rar 0, A5–9, antipet, G usu (5), stigma capitate; ov usu n, free-central. Capsule. *Widespread.*

PLUMBAGINALES

A Antipet; ov 1, basal on long curved stalk.

174. *Plumbaginaceae**†

Herbs/shrubs. Lvs alt/basal, simple, exstip. Infl cymose/racemose/head-like. Fls bisex, act. KCA hypog/CA perig. K(5), C(5)/rar 5, A5, antipet, G(5), stigmas 5; ov 1, basal. Fr indehiscent, retained in K tube. *Widespread.*

EBENALES

Woody; lvs simple; fls act; A usu more numerous than C.

175. *Sapotaceae*†

Woody with milky sap. Lvs alt, simple, exstip, leathery. Infl cymose/fls solit. Fls bisex, act. K hypog, CA perig. K(4–8), C(4–8), A antipet/or n in 2–3 series, staminodes frequent, G1–14-celled; ov 1 per cell, ax. Usu hard berry. *Mostly Trop.*

176. *Ebenaceae*†

Woody. Lvs alt, simple, entire, leathery. Infl cymose/fls solit. Fls usu unisex, act. KCA hypog/K hypog CA perig. K(3–7), C(3–7), A 1–4 × as many as C, G(3–16); ov 1–2 per cell, ax. Berry. *Mostly Trop.*

177. *Styracaceae**†

Woody. Lvs alt, simple, exstip, with stellate hairs or scales. Infl var. Fls bisex, act. KCA hypog/epig/K hypog CA perig. K4–5/(4–5), C(4–5), A8–12 in 1 series, G(3–5); ov n, ax at least below. Drupe dehiscing irregularly. *America, Mediterranean area, E Asia.*

178. *Symplocaceae*

Woody. Lvs alt, simple, exstip, leathery. Infl racemose/fls solit. KCA more or less epig. K(5), C(3–11), A4–n in 1–4 series, G(2–5); ov 2–4 per cell, ax. Berry/drupe. *Warmer Asia, America, Australia (Symplocos).*

OLEALES
Woody; A2.
179. Oleaceae†

Woody, sometimes climbing. Lvs usu opp, simple/pinnately compound, exstip. Infl often cymose panicle. Fls bisex, act. KCA hypog/K hypog CA perig. K(4)/rar (0–15), C(4)/rar (0–15), A2, G(2); ov usu 2 per cell, ax. Fr var. *Temp & Trop.*

GENTIANALES
Lvs usu entire; fls act; A4–5, G(2), sup.
180. Loganiaceae†

Woody, sometimes climbers, with internal phloem; glandular hairs absent. Lvs opp, entire, stip. Infl cymose/fls solit. Fls bisex, act. K hypog, CA perig. K(4–5), imbricate, C(4–5), A4–5, G(2); ov n, ax. Capsule/berry/drupe. *Trop.*

181. Desfontainiaceae

Shrubs. Lvs opp, exstip, evergreen, spiny-margined. Fls solit/cymose, bisex, act. K hypog CA perig. K5, spine-like, C(5) with long tube, A5, G(5); ov n, ax. Berry. *Andes (Desfontainia).*

182. Gentianaceae†

Herbs, rar saprophytic. Lvs opp, entire, exstip. Infl cymose/fls solit, bisex, act. K hypog, CA perig. K4–5/rar –12, usu united, C(4–5)/rar (–12), A4–5/rar –12, G(2); ov n, usu par. Capsule. *Mainly Temp & Subtrop.*

183. Menyanthaceae†

Aquatic/marsh herbs. Lvs alt, entire/of 3 leaflets, stalks sheathing. Infl var. Fls bisex, act. K hypog CA perig, K5/(5), C(5), valvate in bud, A5, G(2); ov n, par. Usu capsule. *Temp.*

184. Apocynaceae†

Woody/herbs, often climbing, with milky sap. Lvs entire, usu opp, exstip. Infl racemose/cymose/fls solit. Fls bisex, act. K hypog CA perig. K4–5/(4–5), C(5), contorted, A5, G(2), often united only by the common style; ov n, marginal/ax. Fr var, seeds often plumed. *Widespread, mainly centred in Trop.*

185. Asclepiadaceae†

Woody/herbs/climbers, usu with milky sap. Lvs opp, entire, stip minute/0. Infl racemose/cymose/fls solit. Fls bisex, act. K hypog, CA perig. K5/(5), C(5), contorted, corona frequent, A5, often joined to

style, 'translators' and pollinia frequent, G(2), often united only by style; ov n, marginal. Fr 1–2 follicles, seeds plumed. *Mostly Trop.*
186. Rubiaceae*†
Woody/herbaceous. Lvs opp, stip (stipules sometimes leaf-like so that the leaves appear whorled). Infl cymose/headlike/fls solit. Fls bisex, usu act. KCA usu epig. K4–5/(4–5), C(4–5) rar (–10), A4–5 rar 10, borne on C tube, G(2 or more); ov 1–n per cell, ax. Capsule/berry/2 mericarps. *Widespread.*

SCROPHULARIALES
C often 2-lipped; A4/2; G(2–3), sup.
187. Polemoniaceae*†
Herbs/rar shrubs or climbers with tendrils. Lvs alt/opp, entire/ pinnately divided, usu exstip. Infl cymose to head-like, rar fls solit. Fls bisex, usu act. K hypog CA perig. K(5), C(5), contorted, A5, G(3); ov 1–n per cell, ax. Capsule. *Mainly America.*
188. Fouquieriaceae†
Woody, spiny. Lvs alt, simple, exstip, fleshy. Infl terminal panicle. Fls bisex, act. K hypog CA perig. K5, C(5), A10–17, G(3); ov 12–18, par. Capsule. *C & SE North America.*
189. Convolvulaceae*†
Climbers/shrublets, often with milky sap/rar twining parasites without chlorophyll. Lvs alt, simple, exstip, scale-like in parasites. Infl cymose/clustered/fls solit. Fls bisex, act. K hypog CA perig. K4–5/(4–5), C(4–5), contorted, A5, sometimes with scales below their insertion, G(2), sometimes 2–4-celled; ov 1–2 per cell, ax/par. Capsule/fleshy. *Widespread.*
   The parasitic genus *Cuscuta* is sometimes separated off into the family Cuscutaceae.
190. Hydrophyllaceae†
Usu herbs. Lvs alt/basal/rar opp, entire/divided, exstip. Infl coiled cymes/fls solit. Fls bisex, act. K hypog CA perig. K5/(5) C(5), usu imbricate, A5, G(2); ov n/rar 4, ax/par. Capsule. *Mainly America.*
191. Boraginaceae*†
Herbs/woody. Lvs alt, simple, exstip. Infl often of coiled cymes. Fls bisex, act/rar zyg. K hypog CA perig. K5/(5) C(5), A5, G(2), usually 4-celled by secondary septa; ov 4, side by side, ax. Style terminal or from between the 4 cells. Fr 4/rar 1 nutlets/drupe. *Widespread.*

101

192. Lennoaceae†

Parasitic herbs, chlorophyll 0. Lvs scale-like. Infl spicate/cymose/ head-like. Fls bisex, act. K hypog CA perig. K(6–10), C(5–8), imbricate, A5–8, G(6–15); ov 2 per cell, ax. Fleshy capsule. *SW USA, Mexico.*

193. Verbenaceae*†

Woody/herbaceous. Lvs opp, simple/compound, exstip. Infl var. Fls bisex, zyg. K hypog CA perig. K(5–8), more or less act, C(5), A4/rar 2–5, G2–9-celled, style terminal; ov 1–2 per cell, ax/rar par. Drupe/ berry/rar 4 nutlets. *Mainly Trop.*

194. Callitrichaceae*†

Aquatic herbs. Lvs opp, simple, exstip. Fls solit, axillary, unisex, act, A hypog. P0, A1, G(2), 4-celled by secondary septa; ov 1 per cell, ax. Schizocarp. *Widespread (Callitriche).*

195. Labiatae/Lamiaceae*†

Herbs/shrubs. Lvs opp, aromatic. Infl often verticillate. Fls bisex, zyg. K hypog CA perig. K usu (5), often zyg, C(5) rar (3), 1–2-lipped, A4/2, G(2), 4-celled by secondary septa; style usually from between the 4 cells of the ovary, rarely terminal; ov 1 per cell, ax. Fr 4 nutlets rar fleshy. *Widespread.*

196. Nolanaceae

Herbs/shrublets. Lvs alt, simple, exstip, fleshy. Fls axillary, bisex, act. K hypog CA perig. K(5), C(5) infolded in bud, A5, G(5), lobed; ov few, ax. Schizocarp. *Chile, Peru.*

197. Solanaceae*†

Woody/herbaceous, with internal phloem. Lvs alt, simple/rar pinnatisect, exstip. Infl often cymose/fls solit, often extra-axillary. Fls bisex, act/zyg. K hypog CA perig. K5/(5) C(5) lobes folded/contorted/ valvate, A5 rar 4/2. G usu (2), septum usu oblique, rar with secondary septa; ov n, ax. Berry/capsule. *Widespread.*

198. Buddlejaceae

Woody/rar herbaceous, without internal phloem, often with glandular hairs. Lvs opp/whorled/rar alt, often toothed, stip forming a line uniting bases. Infl var. Fls bisex, act. K hypog CA perig. K(4), C(4), A4, G(2), style 1; ov n, ax. Capsule/berry/drupe. *Mainly Trop E Asia.*

Often placed close to, or included in Loganiaceae (p. 100).

199. Scrophulariaceae*†

Herbs/woody, some half-parasitic, internal phloem absent. Lvs

alt/opp, simple/rar compound. Infl var. Fls bisex, usu zyg. K hypog
CA perig. K(4–5), C(4–5)/rar (–8), lobes imbricate, A4/2/rar 5, G(2)
septum horizontal; ov 1–n, ax, placentae usu simple. Capsule/rar
berry/indehiscent. *Widespread.*

200. Globulariaceae*

Herbs/shrublets. Lvs alt/basal. Infl a bracteate head. Fls bisex, zyg. K
hypog CA perig. K(5), C(4–5), A4, G(2), 1-celled; ov 1, apical. Nut.
Mostly Mediterranean area.

201. Bignoniaceae†

Usu woody/climbing, often with leaf-tendrils. Lvs usu opp,
compound/rar simple. Infl usu cymose. Fls bisex, zyg. K hypog CA
perig. K(5), C(5), A4/rar 2, G(2); ov n, ax/rar par. Usu capsule; seeds
often winged. *Mainly Trop.*

202. Acanthaceae*†

Usu herbs. Lvs opp, simple, often with cystoliths. Infl cymose, often
with conspicuous overlapping bracts. Fls bisex, zyg. K hypog CA
perig. K(4–5), C(5), 2-lipped, A4/2, G(2); ov ax, 2 or more per cell,
often one above the other. Fr usu explosive capsule. *Mainly Trop.*

203. Pedaliaceae†

Herbs, often sticky-hairy. Lvs opp/alt above, simple, exstip. Infl
racemes/axillary cymes/fls solit. Fls bisex, zyg. K hypog CA perig.
K(5) C(5) A4/rar 2, G(2), 2–4-celled; ov 1–n per cell. Capsule, often
2-horned/nut-like. *Trop, South Africa.*

204. Martyniaceae

Herbs/rar shrubby. Lvs opp/sometimes alt above, exstip. Fls in
cymes/solit, bisex, zyg. K hypog CA perig. K(5), C(5), 2-lipped, A4
with 1 staminode, G(2), 2–4-celled; ov n per cell, ax. Capsule, often
horned/drupe. *Subtrop America.*

205. Gesneriaceae*

Herbs/shrubs, some epiphytic. Lvs usu opp/basal, often velvety. Infl
cymose/fls solit. Fls bisex, usu zyg. K hypog CA perig. K(5), C(5)
A4/2/rar 5, G(2); ov n, par, placentae intrusive, bifid. Capsule/berry.
Mostly Trop.

206. Orobanchaceae*†

Parasitic herbs, chlorophyll 0. Lvs alt, scale-like. Fls bisex, zyg. K
hypog CA perig. K(4–5), C(5), A4, G(2/rar 3); ov n, par, usu on 4
placentae. Capsule. *Mainly N Temp.*

Lathraea, often placed in the Scrophulariaceae, is included here.

207. Lentibulariaceae*†

Herbs, mostly insectivorous, some aquatic. Lvs alt/basal, often of 2 forms, elaborated. Infls on scapes, racemes/fls solit. Fls bisex, zyg. K hypog CA more or less perig. K2–5/(2–5), C(5), spurred at base, A2, G(2); ov n, free-central. Capsule. *Widespread.*

208. Myoporaceae

Woody. Lvs usu alt, often with resinous glands. Infl var. Fls bisex, usu zyg. K hypog CA perig. K(5), C(5), A4/rar 5, G(2); ov 4–8, ax. Fr drupe-like. *Scattered, chiefly Australasian.*

209. Phrymaceae†

Herbs. Lvs opp. Infl spike. Fls bisex, deflexed in fr, zyg. K hypog, CA perig. K(5), teeth hooked, C(5), A4, G(2); ov 1, basal. Nut in persistent K. *E Asia, Atlantic N America (Phryma).*

PLANTAGINALES

Fls act, parts in 4s; C hyaline, small; A exserted; G sup.

210. Plantaginaceae*†

Herbs/rar shrublets. Lvs alt/opp/basal. Infl usu spike. Fls uni/bisex, act. K hypog CA perig/KCA hypog. K4/(4), C(3–4), A usu 4, G1–4-celled; ov few. Capsule opening by lid/nut. *Widespread.*

DIPSACALES

Woody/herbaceous; lvs exstip; G inf.

211. Caprifoliaceae*†

Mostly shrubs/climbers. Lvs opp, usu simple/rar pinnate & exstip. Infl often cymose. Fls bisex, act/zyg, often twinned. KCA epig. K5/(5), C usu (5), A4–5, borne on C tube, G(3–5) sometimes only 1 cell fertile; ov 1–n per cell, ax/pendulous. Berry. *Widespread, mainly N Temp.*

212. Adoxaceae*†

Rhizomatous herbs. Lvs opp/basal, compound, exstip. Fls in a head, bisex, act. KCA more or less epig. K(2–3), C(4–6), A4–6, borne on C tube, each split into 2 half-anthered portions, G(3–5); ov 3–5, ax. Drupe. *N Temp (Adoxa).*

213. Valerianaceae*†

Herbs. Lvs opp, simple/dissected, exstip. Infl cymose. Fls bisex, zyg. KCA epig. K late developing, C(5), often saccate/spurred, A1–4 borne on the C tube, G(3), 1 cell fertile; ov 1, pendulous. Cypsela. K often elaborated in fr. *Mainly N Temp, Andes.*

214. Dipsacaceae*

Mostly herbs. Lvs opp, simple/dissected, exstip. Infl involucrate head, rar fls in spiny-bracted whorls. Fls bisex, zyg. KCA epig with cupular involucel. K5–10 or cupular, C(4–5), A4/rar 2, borne on C tube, G(2); ov 1, apical. Cypsela enclosed in involucel. *Old World, centred in Mediterranean area.*

CAMPANULALES

Lvs alt; A5 rar –2, often free from C and convergent; G usu inf.

215. Campanulaceae*†

Herbs, often with milky sap/rar woody. Lvs usu alt, simple, exstip. Infl var. Fls bisex, act/zyg. KCA epig/rar hypog. K5/rar 3–10, C(5)/rar (3–10), valvate, A5/rar 3–10, rarely attached to C tube, G(2–5)/rar (–10); ov n, ax. Capsule/fleshy. *Widespread.*

Species with bilaterally symmetric flowers are sometimes separated off as the family Lobeliaceae.

216. Goodeniaceae

Herbs/shrubs. Lvs usu alt, simple, exstip. Fls bisex, zyg. KCA usu epig. K5/(5), C(5), 1–2-lipped, valvate/infolded, A5, sometimes attached to C tube, G(2), 1–2-celled, stigma sheathed; ov 1–2 per cell, ax/basal. Fr var. *Mainly Australasia.*

217. Brunoniaceae

Herbs. Lvs basal, simple, exstip. Infl a head with bracts. Fls bisex, more or less act. K hypog CA perig. K(5), C(5), valvate, A5, anthers united into tube, G1-celled; ov 1, basal. Nut enclosed in K tube. *Australia (Brunonia).*

218. Stylidiaceae

Herbs. Lvs basal/on the stem, linear, usu exstip. Infl var. Fls uni/bisex, act/zyg. KCA epig. K(5–7), C(5), imbricate, A2, joined to syle, G(2); ov n, ax/par/free-central. Fr usu capsule. *Australasia.*

219. Compositae/Asteraceae*†

Herbs/woody, sometimes with milky sap. Lvs var. exstip. Infl an involucrate head (rar 1-flowered). Fls uni/bisex, act/zyg. KCA epig. K reduced/as pappus, C(5/3), A(5) rar 5, anthers joined in a tube, G(2); ov 1, basal. Cypsela usu with pappus. *Widespread.*

Subclass Monocotyledones

ALISMATALES

Fls act; G sup, carpels free; aquatics.

220. Alismataceae*†

Scapose aquatics without latex. Lvs often broad. Infl usu much branched, rar 1 umbel. Fls uni/bisex, act. K3, C3, A6–n, G6–n, sup; ov 1–2, basal/marginal. Achenes. *Temp & Trop.*

221. Butomaceae*

Aquatics with or without latex. Lvs basal or on the stem. Infl umbel/fls solit, with bracts. Fls bisex, act. P6/K3, C3, persistent or not, A1–n, G6–n, sup; ov n, diffuse-par. Follicles. *Temp Eurasia, Trop.*

Often split into two families:

1a. Leaves linear, latex absent; all perianth-segments petal-like (*Temp Eurasia*) **Butomaceae**

b. Leaves with stalk and blade, latex present; perianth of 3 sepals and 3 petals (*Trop*) **Limnocharitaceae**

222. Hydrocharitaceae*†

Aquatics with at least the flowers usu emerging from water. Lvs var. Fls (rar solit) arranged in a bifid spathe or between 2 opposite bracts, uni/bisex. K3, C3, A1–n, G usu (3–6), inf; ov n, diffuse-par. Capsule/ rar berry-like. *Mainly Trop & warm Temp.*

223. Scheuchzeriaceae*†

Bog plant, herbaceous. Leaves in 2 ranks with sheaths with ligules. Fls in racemes, with bracts, bisex, act. P6, A6, G3–6, sup; ov 2, basal. Follicles. *Cold N Temp (Scheuchzeria).*

224. Aponogetonaceae

Fresh water aquatics. Lvs long-stalked, sheathing. Infl a simple/forked spike. Fls usu bisex. P1–3/0, sometimes petal-like or (when P1) bract-like, A6 or more, G3–6, sup; ov few, basal. Follicles. *Old World, mainly Trop.*

225. Juncaginaceae*†

Usu marsh plants. Lvs basal, sheathing. Infl raceme/spike, without bracts. Fls uni/bisex, act. P6/rar 1 (when sometimes interpreted as a bract), A1/4–6, G(3–6), sup, 1-celled; ov 1 per cell. Fr capsule/ indehiscent and of 2 forms. *Mainly Temp & cold regions; Pacific America.*

226. Potamogetonaceae*†

Submerged or emergent aquatics of fresh, brackish or sea water. Lvs alt/opp, sometimes in 2 ranks, often sheathing, sometimes with ligules or stipule-like sheath-margins. Fls in bractless spikes/on a flattened axis at first enclosed in a leaf-sheath, uni/bisex, act. P0/represented by lobes or scales/4, A1/3–4, sometimes inserted on P claw, G4/1-celled, rarely more, sup, stigmas sometimes dilated; ov 1, basal or apical. Drupe/achene/indehiscent. *Widespread.*

Frequently divided into several segregate families:

1a. Perianth of 4, clawed segments, valvate; fresh water aquatics with bisexual flowers in submerged or emergent spikes; carpels 4, free (*Widespread*) **Potamogetonaceae**

  b. Combination of characters not as above   2

2a. (1) Marine plants with densely fibrous rhizomes (washed up on beaches as fibre-balls); leaves mostly basal, with ligules; flowers in stalked spikes subtended by reduced leaves (*Mediterranean area, Australia*) **Posidoniaceae**

  b. Plants marine or in brackish marshes; leaves in 2 ranks or opposite, without ligules; flowers in 2-flowered spikes or on a flattened axis at first enclosed in a leaf sheath   3

3a. (2) Marine plants; flowers unisexual; stigmas not dilated (*Widespread*) **Zosteraceae**

  b. Plants of brackish marshes; flowers bisexual; stigmas dilated (*Temp & Subtrop*) **Ruppiaceae**

227. Zannichelliaceae*†

Submerged aquatics of fresh/saline water. Lvs alt/opp/whorled, entire. Fls unisex in axillary cymes/solit. P cupular/3 scales/0, A1–3, G1–9, sup/naked, stigmas dilated/2–4-lobed; ov 1, pendulous. Fr indehiscent, stalked. *Widespread.*

228. Najadaceae*†

Submerged aquatics of fresh/saline water. Lvs opp/whorled, entire/toothed. Fls solit at base of branches, unisex. Male: P2-lipped, A1; female: P membranous/0, G1-celled, sup/naked, with 2–4 stigmas; ov 1, basal. Fr indehiscent. *Widespread (Najas).*

LILIALES

P petal-like, usu act; A usu 6/3; G sup/inf; nectaries often present between the septa on the sides of the ovary.

This large group, together with some others, has been extensively re-classified in recent years, with the recognition of many segregate families. A more traditional view is held here as being more useful for identification purposes. Readers requiring further information should consult Dahlgren, R. T. & Clifford, H. T., *The Monocotyledons: a comparative study* (1982) and Dahlgren, R. T., Clifford, H. T. & Yeo, P. F., *The Families of Monocotyledons: structure, evolution and taxonomy* (1985).

229. *Liliaceae**†

Habit diverse: herbs with rhizomes, corms, bulbs, etc/shrubs/ climbers. Lvs basal/on the stem, with main veins usu parallel to margins, sometimes succulent, spiny-margined, rar reduced to scales when cladodes present. Infl usu racemose/umbels/fls solit. Fls bisex/ rar unisex, act/weakly zyg. P usu 6/(6), mostly petal-like, A usu 6/(6), G usu (3), mostly sup; ov 3–n, ax/rar par. Capsule/berry. *Widespread.*

The following segregate families are those most likely to be encountered (and found in earlier literature):

1a. Leaf-stalk bearing 2 tendrils, or leaf surfaces reversed by a twist in the stalk, or cladodes present in the axils of reduced scale-leaves   2

  b. Plants without any of the above characters   4

2a. (1) Ovary inferior; flowers large and showy; fruit a capsule (*C & S America*)   **Alstroemeriaceae**

  b. Ovary superior; flowers small; fruit a berry   3

3a. (2) Plants with ovate, spiny or thread-like cladodes; leaves scale-like or reduced to small spines (*Tropical and temperate regions of the Old World*)   **Asparagaceae**

  b. Shrubs without cladodes; leaves broad, net-veined, their stalks each bearing 2 tendrils   **Smilacaceae**

4a. (1) Perianth whorls markedly dissimilar, with parts in 3s or more; leaves (except for basal scale-leaves) opposite or usually in a whorl at the top of the stem (*N Temperate regions*)

                                 Trilliaceae

  b. Perianth whorls similar, petal-like, parts usually in 3s; leaves not as above   5

5a. (4) Shrubs or woody climbers with scattered stem-leaves;

flowers solitary, usually pendulous and large; placentation
usually parietal; fruit a berry (*S Hemisphere*)   **Philesiaceae**
  b. Usually herbs with rhizomes, corms or bulbs, rarely herbaceous
      or woody climbers; leaves basal (rarely reduced to sheaths) or
      those on the stems spirally arranged or in several whorls, or if
      plant woody and/or with succulent leaves in terminal crowns
      then perianths tubular and sometimes inflated or irregular;
      flowers usually with parts in 3s, rarely 2s; placentation usually
      axile; fruit a capsule or berry   6
6a. (5) Flowers in umbels subtended by 1 or more bracts at the base
      (*Widespread*)   **Alliaceae**
  b. Flowers not in umbels, bracts various, not as above (*Wide-
      spread*)   **Liliaceae**

230. *Agavaceae*†
Often woody. Lvs alt, usu in basal or terminal rosettes, usually
leathery, fibrous or succulent, often very large and persistent for
many years. Infl racemes/panicles/fls solit. Fls usu bisex, act/zyg.
P6/(6), A6, usu borne near top of P tube, G(3), sup/inf/ov n, ax.
Berry/dry and indehiscent. *Mostly America.*

231. *Haemodoraceae*†
Herbs, sap often orange. Lvs mostly basal, often equitant. Infl
cymes/racemes/clusters. Fls bisex, act/weakly zyg. P6/(6), petal-like,
persistent, often densely hairy, A6/3, G(3) sup/inf; ov n, ax. Capsule.
Mainly S Hemisphere; N America.

232. *Amaryllidaceae**†
Herbs with bulbs/rhizomes/corms. Lvs usu basal, alt, often in 2 ranks.
Umbel/fls solit with 1–several bracts enclosing infl in bud and usu
persisting. Fls bisex, act/zyg. P6/(6), petal-like, often with a corona,
A6, anthers sometimes opening by pores, G(3), inf; ov 2–n per cell,
ax. Capsule/berry. *Widespread.*

233. *Hypoxidaceae*†
Herbs with corms/rhizomes. Lvs usu basal, alt, often pleated and
hairy. Fls solit/racemes/heads, bisex, act. P(6) in two series of 3, those
of the outer series usu hairy outside, A6, G(3), inf; ov n per cell, ax.
Capsule/berry. *Scattered.*

234. *Velloziaceae*
Shrubs with forked branches with persistent leaf-bases/woody-based

herbs. Fls solit, terminal on naked stalks in terminal tufts of lvs. P6/(6), petal-like, A6/more in 6 bundles, G(3), inf; ov n, ax. Hard capsule, often with small spines or glandular. *Trop Arabia, Madagascar, Africa, S America.*

235. *Taccaceae*

Herbs with scapes. Lvs broad, often stalked. Fls in umbels with an involucre, inner bracts often dangling. P(6), more or less petal-like, A6, G(3), inf; ov n, par. Berry/capsule. *Trop & China.*

236. *Dioscoreaceae*†*

Climbers with swollen rootstocks, sometimes with stem-tubers. Lvs on the stem, usu alt, stalked, often cordate/palmate. Racemes axillary. Fls unisex, act, small. P6/(6), often greenish, A6/3, free/united, G(3), inf; ov 2 per cell, ax. Capsule/berry. *Mainly Trop & warm Temp.*

237. *Pontederiaceae†*

Aquatic. Lvs with sheathing bases, often stalked. Infl racemose in axil of spathe-like sheath. Fls bisex, act/rar zyg. P(6), petal-like, A usu 6, G(3), sup; ov 3–n, ax/par. Capsule. *Trop & warm Temp.*

238. *Iridaceae*†*

Terrestrial herbs with rhizomes/corms/bulbs. Lvs often equitant. Infl racemose/fls solit. Fls bisex, act/zyg. P6/(6), petal-like, sometimes in 2 dissimilar whorls, A3, G(3) inf with styles often divided; ov n–few, usu ax. Capsule. *Widespread.*

JUNCALES

Fls clustered but not imbricated into spikelets; P6, hyaline/brownish, papery.

239. *Juncaceae*†*

Lvs often basal, spirally arranged, sometimes reduced to sheaths. Stems round in section. Infl cymes/panicles/corymbs/heads. Fls usu bisex, act. P6, A usu 6, pollen in tetrads, G(3), sup; ov 3–n, ax par. Capsule. *Widespread.*

BROMELIALES

Lvs usu in basal rosettes, stiff & channelled; P strongly differentiated; nectaries frequent in the septa of the ovary.

240. *Bromeliaceae†*

Herbs, often epiphytic. Lvs mainly basal, often spiny-margined or with elaborate hairs. Infl usu terminal raceme/panicle, with conspicuous bracts. Fls bisex, usu act, K3 C3/(3), A6, anthers usu versatile, G(3) usu inf; ov n, ax. Berry/capsule. *Mainly Trop America.*

COMMELINALES
K & C strongly differentiated; staminodes present/0.
241. *Commelinaceae*†
Terrestrial herbs. Lvs mostly on the stems, often with closed basal sheaths. Infl. panicle/coiled cyme/fls solit. Fls bisex, act/weakly zyg. K3, C3 rar fewer, A3–6, anthers basifixed, filaments sometimes hairy; staminodes 0–3, G(3), sup; ov few, ax. *Trop & warm Temp.*
242. *Mayacaceae*†
Aquatic. Lvs on the stems, slender, apex 2-toothed. Fls axillary, bisex, act, K3, C3, A3, anthers opening by pores, G(3), sup; ov several, par. *Trop America, Africa (Mayaca).*
243. *Xyridaceae*†
Terrestrial/marsh plants with mostly basal lvs. Infl a head with bracts. Fls bisex, more or less zyg. K usu 3, segments of differing forms, C(3), A3, staminodes 0–3, G(3), sup; ov few–n, par. Capsule. *Mainly Trop.*
244. *Eriocaulaceae**†
Usu marsh plants. Lvs basal/on the stem, sheathing. Fls in heads subtended by involucres, unisex, act/zyg. P usu in 2 series, 4–6/(4–6), hyaline/membranous, A4–6/2–3, sup; ov 1 per cell, ax. *Mainly Trop.*

GRAMINALES
Fls arranged in 2 ranks in spikelets; P reduced to 2/3 scales (lodicules); fr a caryopsis.
245. *Gramineae/Poaceae**†
Herbs/bamboos. Lvs in 2 ranks, sheathing and usu with ligules; stems round in section, internodes usu hollow. Fls compressed between bract (lemma) and bracteole (palea, rar 0), the unit comprising a floret, these arranged in 2 ranks in spikelets subtended by 2/rar 1 empty bracts (glumes). P represented by 2–3 lodicules. A3/rar 2/rar 6 or more, G1-celled, sup, styles 2/rar 3/1; ov 1, usu lateral. Seed fused to pericarp (caryopsis). *Widespread.*

ARECALES
Palms; usu with woody stems; lvs pleated, splitting into leaflets.

246. Palmae/Arecaceae*†

Trees/shrubs/prickly scramblers. Lvs large & pleated, becoming palmately/pinnately divided. Infl fleshy panicles/spikes, often with large, hard, spathe-like bracts. Fls uni/bisex, act. P6/(6), fleshy, A6 or more, G usu (3), sup, 1–3-celled; ov 3; carpels rar free & with 1 ovule. Berry/drupe, sometimes huge. *Mainly Trop.*

ARALES

Spathe & spadix usu present; lvs often stalked & broad, net-veined.

247. Araaceae*†

Herbs/woody climbers, sap often bitter/milky, rar reduced, floating aquatics. Lvs usu stalked and broad, often lobed. Fls minute, stalkless on a spadix enclosed in a conspicuous spathe, uni/bisex. P4–6/(4–6)/0, A2–8, G1–n-celled, sup/naked; ov 1–n/ Usu berry. *Trop (mainly) & Temp.*

248. Lemnaceae*†

Small floating or submerged aquatics, thallus-like, without true roots. Fls unisex, P0, A1–2, G1-celled, naked; ov 1–7. Fr a utricle. Often reproducing by budding. *Widespread.*

PANDANALES

Woody plants with long, stiff lvs; syncarps.

249. Pandanaceae

Dioecious trees/shrubs, often with stilt-roots. Lvs crowded, leathery, keeled, often with spiny margins. Fls in panicles/spadices, unisex. P rudimentary/0, A n, G1-celled, sup/naked; ov 1–n, basal/par. Fr syncarp, units woody/fleshy. *Old World Trop, Hawaii.*

250. Sparganiaceae*†

Aquatic. Fls in unisex spherical heads. P few scales, A3 or more, G1-celled, superior, more or less stalkless; ov 1, apical. Fr drupe-like. *N Temp, Australasia (Sparganium).*

251. Typhaceae*†

Marsh plants. Infl of 2 unisex, dense, superimposed spikes. P threads/scales, A2–5, G1-celled, sup, on a hairy stalk; ov 1, apical. Fr dry. *Widespread (Typha).*

CYPERALES

Fls subtended by membranous bracts arranged spirally/in 2 ranks in spikes/in spikelets; P reduced; fr nut-like.

252. *Cyperaceae**†

Herbs. Stems round/3-sided in section, usu solid. Lvs spirally arranged, with closed sheaths. Fls subtended by membranous bracts (glumes), spiral/in 2 ranks in spikes/spikelets without 2 empty glumes at the base, uni/bisex. P scales/bristles/hairs/0, A6/rar 3, with basi-fixed anthers, G1-celled, sup/naked, sometimes surrounded by a flask-shaped structure, styles 2–3; ov 1, basal. Fr nut-like. *Widespread*.

ZINGIBERALES

Lvs usu stalked with broad blades pinnately parallel-veined; fls zyg/asymmetric; A6–1; G inf often with nectaries on the septa.

253. *Musaceae*

Herbs/shrubs, often very large, sometimes leaf-stalks rolled and forming a stem/trunk-like structure. Lvs spiral/in 2 ranks. Infl coiled cymes/in the axils of spathes. Fls uni/bisex, zyg. K3/(3), sometimes joined to the petals or tubular and split down 1 side, C3/(3)/2-lipped, A5, sometimes with 1 staminode, G(3) inf; ov n/rar 1 per cell, ax. Capsule/schizocarp/fleshy. *Trop*.

Often divided into segregate families:

1a. Leaves and bracts spirally arranged; flowers unisexual; fruit a banana (*Old World Tropics*) **Musaceae**

 b. Leaves and bracts in 2 ranks; flowers bisexual; fruit not a banana 2

2a. (1) Cymes arising from the bases of leaf-sheaths; sepals united below into a long, stalk-like tube; median (upper) petal forming a large lip (*Trop SE Asia*) **Lowiaceae**

 b. Flowers in coiled cymes in the axils of spathes; sepals free or at most joined to corolla; petals not forming a lip 3

3a. (2) Perianth-segments free; ovary with numerous ovules in each cell (*Trop America, South Africa, Madagascar*)

Strelitziaceae

 b. Perianth-segments partially united; ovary with 1 ovule per cell (*Trop America*) **Heliconiaceae**

254. Zingiberaceae

Lvs in 2 ranks, usu with open sheaths and ligules, aromatic. Fls in racemes/heads/cymes, bisex, zyg. K(3), C3/(3), A1, staminodes usu petal-like, stamen modified into a lip, lateral staminodes present/absent, G(3), inf, usu surmounted by epig glands, style often supported in a groove in the anther; ov n, ax/par. Fr usu capsule. *Trop.*

255. Cannaceae†

Lvs spirally arranged, without ligules. Infl terminal with fls in pairs. K3, C(3), A1, petal-like with a half-anther, with petal-like staminodes, G(3), inf, style petal-like; ox n, ax. Warty capsule. *Trop America (Canna).*

256. Marantaceae†

Lvs in 2 ranks, stalk with a swollen band (pulvinus) at apex. Infl panicle/spike with asymmetric flowers in pairs. K3, C(3), A1 with various petal-like staminodes, G(3), inf; ov. 1 per cell (2 sometimes aborting). Capsule, often fleshy. *Trop.*

ORCHIDALES

Fls zyg with 1 petal usu forming a lip (labellum); A1–2; G inf; seeds with undifferentiated embryo.

257. Orchidaceae*†

Terrestrial/epiphytic/saprophytic herbs. Lvs alt/rar opp. Fls in racemes/solit, bisex, zyg. K3 rar 2/(3), P3, all petal-like, median petal usu modified into a lip of varying complexity, A usu 1–2, joined to style to form the column, pollinia usual, G(3), inf, usu twisted in flower (unless infl pendulous); ov n, par. Capsule. *Widespread.*

Further identification and annotated bibliography

The identification of the family to which a plant belongs is only the first necessary step in its complete identification. To make the book more generally useful we have provided below some notes on the more important literature which can be used for identification. Three broad situations can be defined in so far as identification is concerned. These require somewhat different approaches, and are dealt with separately below. Books and papers are referred to by numbers, and are listed numerically (and alphabetically by author) at the end of the chapter.

1. The family cannot be satisfactorily identified using the present key.

This key does not include all currently recognised flowering plant families. All exclusively tropical and southern hemisphere families have been excluded unless they contain plants cultivated in the northern hemisphere; when this is the case, only those plants cultivated are covered by the key, and other members of these families may well not key out. If the family cannot be identified here, several other works may be used. Some of these have keys (5, 12, 13, 15, 23), others are descriptive, often with illustrations (4, 9, 14, 22, 26, 28).

It is important to emphasise here that the circumscription of a particular family can vary from book to book. Care must be taken, therefore, to see that the family arrived at in this book (or any other book) corresponds closely with the family of the same name in another work. This is often extremely difficult; it may, however, be done by checking the indexes, descriptions, synonyms and comments in the various works against one another. This *caveat* applies to all the works mentioned in this section.

2. The specimen has been identified to family and its wild, geographical origin is known.

When the geographical origin of a specimen is known, further identification may be attempted using a Flora of the region or country in question, if one exists. Floras are too numerous to list here, but details can be found in 3 & 11. If no relevant Flora exists, the specimen must be treated as though its origin was not known, as below.

3. The specimen has been identified to family but its wild geographical origin is not known.

Under these circumstances, the first step is to find out if a world-wide monographic study of the family exists. The most notable series of such monographs is that edited by Engler and his successors (8), but this is by no means complete. Other monographic studies are published from time to time in various botanical journals. Most botanical libraries maintain lists of such publications, and the *Kew Record* (19) publishes short abstracts of current publications.

Attempts to identify the genus to which a plant belongs can be made using various works: 4, 9, 15, 16, 23, which all attempt to be comprehensive (even though some of them are incomplete).

If the specimen is from a garden plant, then it may be possible to identify it using a garden Flora; several of these exist: 2, 7, 20, 21, 27, 29, 32.

It is often helpful to confirm an identification by comparison with a good illustration. Lists of these can be found in 17 & 31. In recent years, many popular illustrated works on both wild and garden plants have been produced which, though selective and botanically simplified, can be helpful; volume 3 of 32 contains a bibliography in which many such works are listed.

There are several other useful works which, though not in themselves usable for identification, contain much useful information. Such works include botanical glossaries and dictionaries (1, 6, 10, 18, 24, 25).

Finally, the value of comparing the identified specimen with named herbarium specimens cannot be overemphasised. This is the most stringent test of the accuracy of an identification, although the warning should be given that herbarium material is sometimes wrongly named. Use of the herbarium is also helpful when the specimen to be named is too incomplete for identification by means of keys.

Annotated bibliography

1. Airy-Shaw, H. K. (ed.), *J. C. Willis, A dictionary of the flowering plants and ferns*, 8th edn, 1973. A valuable source for plant names, and, especially, indication of which genera belong to which families.

2. Bailey, L. H., *A manual of cultivated plants*, 2nd edn, 1949. A detailed account, with keys, descriptions and illustrations, of the 5,000 or so species most commonly cultivated in North American gardens.

3. Blake, S. F. A. & Atwood, A. C., *Geographical guide to the Floras of the world:* Part 1, Africa, Australasia, N & S America and islands of the Atlantic, Pacific and Indian Oceans, 1942; part 2, western Europe, 1961. Lists Floras of the areas mentioned up to the dates specified.

4. Bentham, G. & Hooker, J. D., *Genera Plantarum*, 1862–83. In Latin. Now very out-of-date but still useful for its synopses of genera in each family and its very fine descriptions.

5. Cronquist, A., *An integrated system of classification of flowering plants*, 1981. With excellent descriptions and illustrations of the various families; the keys provided are synoptic, that is, they do not take account of all exceptions.

6. Davidov, N. N., *Botanicheskii Slovar'*, 1960. A multilingual botanical dictionary (Russian, English, German, French, Latin).

7. Encke, F. (ed.), *Parey's Blumengärtnerei*, 2nd edn, 1958. In German. A taxonomic account, with keys to families and genera, descriptions and illustrations of plants widely cultivated in Germany.

8. Engler, A. *et al.* (eds.), *Das Pflanzenreich*, 107 volumes, 1900–53. A series of family monographs with keys, descriptions and illustrations; incomplete.

9. Engler, A. & Prantl, K., *Die Natürlichen Pflanzenfamilien*, several volumes, 1887–99; 2nd edn, several volumes, incomplete, 1924 onwards. In German. With keys, descriptions and illustrations.

10. Featherley, H. I., *Taxonomic terminology of the higher plants*, 1959, facsimile edn, 1965. A standard taxonomic glossary.

11. Frodin, D. G., *Guide to standard Floras of the world*, 1984. The most up-to-date listing of Floras.

12. Geesinck, R., Leeuwenburg, A. J. M., Ridsdale, C. E. & Veldkamp, J. F., *Thonner's analytical key to the families of flowering plants*, 1981. A very full and complete key; uses a different taxonomic system and very different terminology from this book; the 'Introduction and notes' should be read carefully before attempting use of the key.

13. Hansen, B. & Rahn, K., Determination of Angiosperm families by means of a punched-card system, *Dansk Botanik Arkiv* 26(1), 1969. A key to families using easily sorted punched cards; the introduction should be carefully read before use.

14. Heywood, V. H. (ed.), *Flowering plants of the world*, 1978 and subsequent reprints. With descriptions, distribution maps and beautiful illustrations for the families, but no keys.

15. Hutchinson, J., *The families of flowering plants*, 2nd edn, 1959. With keys to families (and sometimes to the genera within them), descriptions and some illustrations; follows an idiosyncratic taxonomic system.

16. Hutchinson, J., *The genera of flowering plants*, vol. 1, 1964, vol. 2, 1967. Only 2 volumes completed; with keys and descriptions for the genera covered.

17. Isaacson, R. T., *Flowering plant index of illustration and information*, 2 volumes, 1969. References to recent illustrations and articles relevant to garden plants.

18. Jackson, B. D., *A glossary of botanic terms*, 4th edn, reprinted 1953. A standard botanical glossary.

19. *Kew Record of Taxonomic Literature*, 1971 and continuing. Contains abstracts of books and articles of taxonomic interest arranged by families, genera and species, etc.

20. Kirk, W. J. C., *A British garden Flora*, 1927. With keys, descriptions and some illustrations to families and genera of plants cultivated in Britain.

21. Krüssmann, G., *Handbuch der Laubgeholze*, 1962, translated into English by Epps, M. as *Manual of cultivated broad-leaved trees and shrubs*, 1986. A very full account with some keys, descriptions and illustrations of woody plants cultivated in Europe.

22. Lawrence, G. H. M., *Taxonomy of vascular plants*, 1951. With descriptions and illustrations of the families, with much other

information, especially references to monographs; without keys.

23. Lemée, A. *Dictionnaire descriptif et synonymique des genres de plantes phanérogames*, 8 volumes, 1925–43. In French. With detailed descriptions of families and genera; keys in volumes 8a and 8b.

24. Mabberley, D. J., *The plant-book*, 1987. A dictionary of information about plants, with brief descriptions, notes and indication for each genus of the family it belongs to.

25. Melchior, H. (ed.), *Syllabus der Pflanzenfamilien*, 12th edn, 1964. In German. With good descriptions and illustrations of the families, but no keys. Contains much other matter, including lists of important genera in each family, division of the families into Subfamilies, Tribes, etc.

26. Nijdam, J., *Woordenlijst voor de Tuinbouw in zeven talen*, 1952. A polyglot horticultural/botanical dictionary (Dutch, English, French, German, Danish, Swedish and Spanish).

27. Rehder, A., *Manual of cultivated trees and shrubs*, 2nd edn, 1949. A very detailed account with keys and descriptions of woody plants cultivated in North America.

28. Rendle, A. B., *Classification of flowering plants*, vol. 1, 1930, vol. 2, 1938. With descriptions and illustrations of the families, but no keys; numerous notes on structures of interest.

29. *Royal Horticultural Society's Dictionary of Gardening & supplement*, 1951–56. A dictionary treatment of plants in cultivation in Britain; some illustrations, very brief descriptions, keys to the species of some genera.

30. Schneider, C. K., *Illustriertes Handbuch der Laubholzkunde*, 1904–12. In German. A well-illustrated, very detailed account, with keys, of woody plants cultivated in Europe.

31. Stapf, O. (ed.), *Index Londinensis*, 4 volumes and supplement, 1921–41. A very complete listing of plant illustrations up to 1941.

32. Walters, S. M. *et al.*, *The European Garden Flora*, vol. 1, 1986, vol. 2, 1984, vol. 3, 1989. A very full treatment with keys, descriptions and some illustrations of all plants widely cultivated in Europe, about half-completed. Vol. 3 contains a bibliography which includes references to many popular illustrated works on plants.

Glossary

Only very brief definitions are given here; if more detail is required, reference should be made to the glossaries cited on pp. 117–119, or to a botanical textbook.

achene: a small, dry, indehiscent 1-seeded fruit; in the strict sense, such a fruit formed from a free carpel.

actinomorphic: regular, radially symmetric, having 2 or more planes of symmetry (see p. 21).

aestivation: the manner in which the perianth parts are arranged relative to one another in the bud (see p. 18).

androecium: the male sex organs of a single flower (stamens) collectively.

androgynophore: the stalk which bears the ovary and stamens in *Passiflora*.

anther: that part of the stamen in which the pollen is produced.

antipetalous: generally of stamens when they are of the same number as, and on the same radii as the petals.

antisepalous: (stamens) as for antipetalous, but on the same radii as the sepals.

apetalous: lacking a corolla.

apical: (placentation) see p. 18.

apocarpous: having free carpels.

aril: an appendage borne on the seed, strictly an outgrowth of the funicle.

asymmetric: (corolla) having no planes of symmetry.

axile: (placentation) see p. 15.

axillary: in the axil (the angle between leaf-stalk and stem).

basal: (placentation) see p. 18.

basifixed: attached by the base.

berry: a fleshy, indehiscent fruit with the seeds immersed in pulp.

bifid: divided into 2 shallow segments.

bilabiate: 2-lipped.

bipinnate: (leaf) a pinnately divided leaf with the segments themselves pinnately divided.

bract: a frequently leaf-like organ (often very reduced) bearing a flower, inflorescence or partial inflorescence in its axil.

bracteole: a bract-like organ (often even more reduced) borne on the flower-stalk.

bulb: an underground organ composed of stem and swollen leaf-bases, enclosing the bud for the next year's growth.

caducous: falling off early.

calyptrate: (usually of perianth) shed as a unit, often in the shape of a cap or candle-snuffer.

calyx: the outer whorl(s) of the perianth, consisting of sepals.

capitate: head-like.

capsule: a dehiscent, usually dry fruit formed from an ovary of united carpels.

carpel: the organ containing the ovules; when several are united they may be much modified and difficult to distinguish.

caruncle: an outgrowth near the point of attachment (hilum) of a seed.

caryopsis: an achene with the seed united to the fruit wall.

catkin: a unisexual inflorescence of small flowers with no petals, often deciduous as a whole and with overlapping bracts.

caudex: the often woody zone joining root and stem.

cladode: a lateral, usually flattened stem-structure borne in the axil of a reduced leaf.

claw: the conspicuously narrowed base of an organ, especially a petal.

compound: (leaf) divided into distinct and separate leaflets.

connective: the part of the stamen which joins the anther-cells.

contorted: see p. 19.

cordate: heart-shaped.

corona: an outgrowth, usually petal-like, of the corolla, stamens or staminodes.

corymb: a flat-topped inflorescence in which branches arise at different levels.

cotyledon: the first seedling leaf.

cupule: a cup formed from free or united bracts, often containing an ovary.

cyme: a determinate or centrifugal inflorescence.

cypsela: a small, indehiscent, dry, 1-seeded fruit formed from an inferior ovary; often loosely termed an achene.

cystolith: a mineral concretion found in special cells in certain leaves, perceptible usually by feel.

deciduous: (leaves) falling once a year; also used of stipules, catkins, etc.

dehiscence: the mode of opening of an organ, usually an anther or fruit.

determinate: (inflorescence) one in which the terminal flower opens first and stops further growth of the axis.

diffuse-parietal: (placentation) see p. 15.

dioecious: with male and female flowers on separate plants.

disc: a fleshy, nectar-secreting organ frequently developed between the stamens and ovary (sometimes also extending outside the stamens).

divided: (leaves) see p. 20.

dorsifixed: attached to its stalk or supporting organ by its back, usually near the middle.

drupe: a fleshy or leathery 1–few-seeded fruit with a hard inner wall.

endosperm: storage material in many seeds, formed after fertilisation.

entire: (leaves) simple and with smooth margins.

epicalyx: a whorl of sepal-like segments borne outside the true calyx.

epidermis: the outermost cell-layer of plants.

epigynous: see pp. 4–10.

epiphytic: growing on another plant but not parasitic.

equitant: (leaves) folded sharply inwards from the midrib, the outermost leaf enclosing the next, etc.

exfoliating: (bark) scaling off in large flakes.

exstipulate: without stipules.

extrorse: (anthers) opening towards the outside of the flower.

false septum: a secondary septum in an ovary, formed after the formation of the primary septa.

filament: the stalk of a stamen, bearing the anther.

-foliolate: divided into a specified number of leaflets (e.g. 3-foliolate).

follicle: a several-seeded fruit or partial fruit formed from a single carpel, dehiscing along one suture.

free-central: (placentation) see p. 18.

fruit: the structure(s) containing all the seeds produced by a single flower.

funicle: the stalk connecting the ovule (and seed) to the placenta.

gamopetalous: with the corolla-segments (petals) united at the base.

gland: a secretory organ.

glumes: see p. 111.

gynoecium: the female sex organs of a single flower collectively.

gynophore: the stalk of a stalked ovary.

halophytic: growing in saline soils.

hemiparasitic: partial parasites with green leaves.

herb: applied to any plant dying back more or less to ground level during each unfavourable season; including plants slightly woody at the base, and annuals.

hypogynous: see p. 4.

imbricate: see p. 19.

indehiscent: (fruit) not opening by any definite mechanism.

indeterminate: (inflorescence) one in which the lower or outer flowers open first and the axis continues growth.

inferior: (ovary) see p. 5.

internode: that part of a stem between one leaf-base and the next.

introrse: (anthers) opening towards the inside of the flower.

intrusive: (placenta) see p. 15.

involucel: a cup-like structure surrounding the inferior ovary of some Dipsacaceae.

involucre: a series of bracts (often overlapping) surrounding an inflorescence.

labellum: a lip, generally used of the enlarged lower petal of some monocotyledonous flowers.

laciniate: deeply slashed into narrow divisions.

leaflet: the separate parts into which compound leaves are divided; distinguished from leaves by their not having buds in their axils.

legume: a dry, dehiscent fruit formed from a single carpel, dehiscent along both sutures.

lemma: see p. 111.

lepidote: bearing peltate scales; or such scales themselves.

ligule: a tongue-like outgrowth from a petal or at the junction of leaf-sheath and blade.

loculicidal: dehiscence of a capsule down the centres of the cells.

locules: the cavity(ies) in a carpel, ovary or anther.

lodicules: see p. 111.

lomentum: an indehiscent fruit which fragments transversely between the seeds.

marginal: (placentation) see p. 15.

medifixed: (hairs, anthers) attached by the middle.

mericarp: a 1-seeded portion of a fruit formed from an ovary of united carpels which splits apart at maturity.

-merous: indicating the numbers of parts (e.g. 3-merous or tri-merous).

monoecious: with plants of different sexes on the same plant.

naked: (ovary) see p. 9.

nut: a hard, indehiscent, 1-seeded fruit.

nutlet: a nut-like, partial mericarp.

obconical: top-shaped.

obdiplostemony: having the stamens twice as many as the petals, those of the outer whorl on the same radii as the petals.

ovules: the structures in the ovary which become seeds after fertilisation.

palea: see p. 111.

palmate: (leaves) divided to the base into separate leaflets, all the leaflets arising from the apex of the stalk.

panicle: a much branched inflorescence.

pappus: a ring of hairs or scales on the top of a cypsela; thought to be a modified calyx.

parietal: (placentation) see p. 15.

partly inferior: (ovary) see p. 5.

peltate: disc-shaped, the stalk arising from the under-surface.

perianth: the outer, sterile whorls of a flower, often differentiated into calyx and corolla.

pericarp: the fruit wall.

perisperm: storage tissue in some seeds formed from maternal tissue.

petal: a single segment of the corolla.

phyllodic: (leaves) having flattened, leaf-like stalks instead of a true blade.

pinnate: (leaf) bearing separate leaflets along each side of a common stalk.

pinnatisect: pinnately divided almost to the midrib.

pistillode: a rudimentary, non-functioning gynoecium.

placenta: that part of the carpel that bears the ovules.

pollinia: coherent masses of pollen dispersed as units.

polypetalous: with distinct, free petals.

pome: a fleshy fruit in which the carpels are immersed in flesh formed from the torus.

poricidal: (anthers) opening by pores.

punctate: dot-like.

raceme: a simple, elongate, indeterminate inflorescence with stalked flowers.

radical: (leaves) arising directly from the rootstock.

radicle: the part of the embryo which grows into the root.

receptacle: the top of a flower-stalk, bearing the floral parts (see torus).

rhizomatous: having rhizomes, i.e. underground stems bearing scale-leaves and adventitious roots.

saccate: (perianth or corolla) with a conspicuous, hollow swelling.

sagittate: arrow-head shaped.

samara: a dry, winged, indehiscent fruit or mericarp, usually 1-seeded.

saprophyte: a plant which obtains its food materials by absorption of complex organic chemicals from the soil; often without chlorophyll.

scape: a leafless flower-stalk arising directly from a rosette of basal leaves.

schizocarp: a fruit which splits into separate mericarps.

sepal: a segment of the calyx.

septicidal: a capsule which splits open through the septa.

septum: the internal partitions of an ovary or fruit.

-seriate: in a number of series, e.g. 3-seriate.

serrate: with sharp, more or less saw-like teeth.

sessile: not stalked.

simple: (leaves) not divided into separate leaflets (but possibly lobed).

spadix: a fleshy spike of numerous small flowers.

spathe: a large bract sheathing an inflorescence.

spike: a raceme-like inflorescence in which all the flowers are stalkless.

spikelet: a secondary spike, a group of 1 or more flowers subtended by bracts.

spur: (of perianth or corolla) a long, usually nectar-secreting tubular projection.

stamen: the male sex organ, usually consisting of filament, anther and connective.

staminode: a sterile stamen.

stellate: (hair) star-shaped.

stigma: the receptive part of the ovary, on which the pollen germinates.

stipulate: with stipules.

stipules: a pair of lateral outgrowths arising at the base of the leaf-stalk.

style: the often elongate portion of the ovary, bearing the stigma(s) at its tip.

superior: (ovary) see p. 4.

superposed: one above the other.

syncarp: a multiple or aggregate fruit, often fleshy or woody.

syncarpous: (ovary) with the carpels united.

tendril: a touch-sensitive, thread-like organ coiling around objects touched.

tetrads: groups of 4 pollen grains shed as units.

torus: the receptacle (q.v.).

translator: a specialised structure uniting the pollinia in most Asclepiadaceae.

trifid: shortly divided into 3.

trifoliolate: (leaves) divided into 3 leaflets.

triquetrous: 3-sided.

truncate: ending abruptly, as though broken or cut off.

umbel: a usually flat-topped inflorescence in which all the flowerstalks arise from the same point in the inflorescence.

utricle: a bladdery, indehiscent, 1-seeded fruit.

valvate: see p. 19.

versatile: (anthers) pivoting freely on the filament.

verticillate: (inflorescence) the flowers in superposed whorls, each whorl consisting of 2 opposite, often modified cymes.

xeromorphic: with the habit of plants characteristic of arid regions – e.g. with reduced or fleshy leaves, or densely hair, etc.

zygomorphic: bilaterally symmetric, having only 1 plane of symmetry.

Index

This index includes only references to the occurrence of family names in the key, and the occurrence of order and family names in the 'Family Descriptions' section.